Modeling of Turbomachines for Control and Diagnostic Applications

Edited by Igor Loboda and Sergiy Yepifanov

Published in London, United Kingdom

IntechOpen

Supporting open minds since 2005

Modeling of Turbomachines for Control and Diagnostic Applications
http://dx.doi.org/10.5772/intechopen.81984
Edited by Igor Loboda and Sergiy Yepifanov

Contributors
Imane Idrissi, Houcine Chafouk, Rachid El Bachtiri, Maha Khanfara, Cristhian Herrera Maravilla, Sergiy Yepifanov, Felix Sirenko, Roman Zelenskyi, Igor Loboda

First published in London, United Kingdom, 2020 by IntechOpen
IntechOpen is the global imprint of INTECHOPEN LIMITED, registered in England and Wales, registration number: 11086078, 7th floor, 10 Lower Thames Street, London, EC3R 6AF, United Kingdom
Printed in Croatia

British Library Cataloguing-in-Publication Data
A catalogue record for this book is available from the British Library

Additional hard and PDF copies can be obtained from orders@intechopen.com

Modeling of Turbomachines for Control and Diagnostic Applications
Edited by Igor Loboda and Sergiy Yepifanov
p. cm.
Print ISBN 978-1-78984-650-8
Online ISBN 978-1-78984-651-5
eBook (PDF) ISBN 978-1-83880-620-0

For EU product safety concerns:
IN TECH d.o.o., Prolaz Marije Krucifikse Kozulić 3, 51000 Rijeka, Croatia, info@intechopen.com or visit our website at intechopen.com.

Meet the editors

Igor Loboda received his MS and PhD degrees in Aircraft Engine Engineering from the Kharkov Aviation Institute (Ukraine) in 1979 and 1994, respectively. He was an investigator, lecturer, and assistant professor at the Kharkov Aviation Institute from 1992 to 2001. Since 2001 he has been an assistant professor and investigator at the National Polytechnic Institute of Mexico. His research interests are in the areas of simulation and condition monitoring of gas turbines and the common theory of pattern recognition. Particular issues of interest are gas turbine thermodynamic models (static and dynamic), model identification, analysis of real data (gas path variables), fault identification techniques, and neural network application to gas path diagnostics.

Professor Sergiy Yepifanov is a Professor (2003) and a Doctor of Sc., Eng. (2001) and the Head of the Aircraft Engine Design Department of the National Aerospace University "Kharkiv Aviation Institute" (Ukraine). He is an Honored Worker of Science of Ukraine. He has supervised 10 PhD dissertations. He is the co-author of 7 monographs and textbooks, as well as 33 scientific articles in indexed journals (SCOPUS) and more than 300 other publications. He is the chief organizer of the Engine Engineering Congress (every year since 1996). He is the lead lecturer of the education disciplines "Construction of Aircraft Engines", "Dynamics and Strength of Aircraft Engines", "Automatic Control Systems of Aircraft Engines", "Diagnostics of Aircraft Engines", "On-ground Application of Aircraft Turbine Engines", "Cooling Systems and Thermal Strength of Aircraft Turbine Engines", and "Life-time Designing of Aircraft Turbine Engines". His research interests are in the areas of turbine engine simulation, automatic control, and health management. He supervised and participated in creation of onboard and on-ground engine diagnosing systems of several aircrafts and engines that were developed in Ukraine and Russia.

Contents

Preface

This book is devoted to the issue of modeling and simulation of turbomachines, namely gas turbines and wind turbines. Gas turbines are the main engines for power generation and transportation, and wind turbines present an increasingly growing sector of renewable energy production. The design of these complex machines and, in particular, the development of control and diagnostic systems rely on mathematical modeling.

The book consists of three sections that include five chapters. Section 1 consists of the introductory chapter (Chapter 1), which introduces the area of modeling of gas and wind turbines and explains the demand for further model development.

Section 2 unites three chapters that offer particular improvements in gas turbine modeling. A novel methodology for the modeling of engine starting is presented in Chapter 2. Chapter 3 performs a thorough theoretical comparative analysis of models for the simulation of gas capacities between engine components, and offers practical recommendations on model applications, in particular for engine control purposes. In Chapter 4, the reader interested in gas turbine diagnostics can find the answer to how to compute important unmeasured parameters. Multiple algorithms for calculating these parameters are proposed and compared. The best algorithms allow accurate prognostics of engine lifetime.

In section 3, the field of wind turbine modeling is presented in Chapter 5. It introduces a general-purpose model that describes both aerodynamic and electric parts of a wind power plant. Such a detailed physics-based model helps with the development of more accurate control and diagnostic systems that will result in more effective electricity production.

In this way, the book includes four new studies in the area of gas and wind turbine modeling. These studies are interesting and useful for specialists in turbine engine control and diagnostics.

Igor Loboda
Instituto Politécnico Nacional,
Ciudad de México,
México

Sergiy Yepifanov
National Aerospace University – Kharkiv Aviation Institute,
Kharkiv, Ukraine

Section 1

Introduction

Chapter 1

Introductory Chapter: Gas and Wind Turbines and Their Models

Igor Loboda

1. Introduction

Turbomachines, in particular, a gas turbine (GT) and a wind turbine that are considered in the present book, are indispensable for power generation, transportation, and many other industries.

In the last decades, a GT industry has exposed a considerable growth [1]. A gas turbine combines high power and efficiency with relatively low weight and compactness. A turboshaft engine presents a principal driver for electricity production. It is also widely used in offshore oil platforms, petrochemical plants, refineries, gas stations, and so on. Many marine and other kinds of transport are equipped by gas turbine engines as well. Depending on a particular application and the range of produced power, the types of GTs embrace heavy-duty, aircraft-derivative, industrial, vehicular, small, and micro gas turbines [1]. Another large area of GT application is air transport. Different aircraft gas turbine engines, which can generally be divided into turbofans, turboprops, and turboshafts, have revolutionized the aviation industry [2] and so far have no alternatives.

Following a general development of the industry of gas turbines, the use of their mathematical models becomes more intensive. Modeling and simulation of GTs have been an effective way of design and manufacture. For example, the use of engine models enables to evaluate and optimize the engine performance before the engine fabrication. In addition to the design of the gas turbine engine itself, engine modeling and simulation are widely used in the development of control and monitoring systems [2–4].

To have high power and efficiency performances, a gas turbine must operate near its functional and structural limits at different steady-state and dynamic modes. For doing so, an engine control system should be accurate enough. In the beginning of the gas turbine technology development, much experimentation was done to design this system. However, since the engine and its control system are more and more complex and expensive, physical experimenting to tailor, test, and validate the system is expensive and therefore limited. For this reason, mathematical modeling and simulation of GTs have been increasingly useful and effective methodology of the control system development. As a result of investigations, better understanding of engine dynamic behavior was gained, and more complex and exact engine dynamic models were developed. These models allow accurate predicting and analyzing engine dynamics. They also help a lot with decreasing fuel consumption and the development and optimization of multiple engine control laws.

The abovementioned complexity and expensiveness of GTs and high demands to engine reliability and safety at low maintenance costs make it unavoidable to use advanced monitoring systems. Through diagnostic processing of engine measured parameters, these systems allow us to gain the knowledge about the health conditions of main engine components and subsystems. Ideally, it is possible to have the necessary information about the influence of engine faults on the measured parameters by physical embedding the faults in real engines. However, it is a too risky and expensive way, and the assessment of engine technical state is usually based on different mathematical models. Since the diagnostic quality strongly depends on the accuracy of the models used, the issue of the diagnostic models development is important [4]. Although, many diagnostic GT models have been proposed and constructed so far, the existing demand for optimized models for different objectives and engine applications motivates the investigators to continue their efforts to further enhance gas turbine modeling.

Let us now come back to wind turbines. They present the principal source of renewable energy. Although wind turbines now generate only about 1% of the total energy, their development rate is by far higher than the rates of traditional energy sources, and the total power production is already about two times greater than the production of solar plants [5]. Being a megastructure, a wind turbine must meet strict requirements of reliability and safety. On the other hand, massive electricity production should be cost-effective. Due to these contradictory demands, as with gas turbines, the turbine itself and its control and monitoring systems need thorough optimization. To this end, a wide use of accurate mathematical modeling and simulation is a very effective strategy. For specific purposes, many different wind turbine models, e.g., aerodynamic, mechanical, economical, and environmental, have been proposed so far. For the needs of control and diagnostics, the general-purpose model that describes both aerodynamic and electric parts of the power plant is mainly used.

2. Gas turbine modeling

As with the models of other technical systems, the gas turbine models fall into two general categories: physics-based models and data-driven models. The physics-based models rely on the theory of gas turbines and therefore perform accurate simulation of various steady-state and transient modes. Being more complex than the data-driven models, physics-based models have however extended capabilities and offer the knowledge that can hardly be obtained from real data, for example, the information about the influence of the faults on engine performances. The data-driven models (aka black box models) do not require the information about internal structure and operation of the simulated engine. To build these models, optimization methods or, in the case of neural networks, machine learning techniques are employed using available real information as input data. These models are widespread because of their simplicity. The variety of such models is described in detail in the book [3].

2.1 Thermodynamic model and its derivations

Among physics-based gas turbine models, a nonlinear component-based model also called a thermodynamic model is primarily used for development of control and diagnostic systems. The model relies on the gas turbine theory [6]. The model involves aerothermal relations to compute gas path variables, and engine components such as compressors, combustion chamber, and turbines are given by their performance maps. The most of diagnostic methods are applied to engine at steady states. In this case, model output variables $\vec{Y}$ are computed employing operational conditions $\vec{U}$ and health parameters $\vec{\Theta}$ as input data. In this way, the static model structure is given by

$$\vec{Y} = F\left(\vec{U}, \vec{\Theta}\right). \tag{1}$$

The health parameters can shift a little the components' maps that allow considering a varying technical state of the engine. Mathematically, expression (1) is a result of solving a system of nonlinear algebraic equations that reflect the balance of mass, heat, and mechanical energy. For the engines of complex structure, the number of equations reaches 15–20.

The development of control algorithms and some diagnostic methods (see, e.g., [7]) relies on dynamic simulation. Within the thermodynamic model, the simulation at transients can be expressed by

$$\vec{Y} = F\left(\vec{U}(t), \vec{\Theta}, t\right). \tag{2}$$

In contrast to the static simulation according to Eq. (1), the operating conditions $\vec{U}$ are now time functions. Time t is also separately added to the arguments to take into consideration inertia factors, namely, inertia moments of engine rotors, mass and energy accumulation in gas capacities, and warming-up of massive metal elements. Expression (2) is determined by the solution of differential equations that reflect combined operation of the engine components under dynamic conditions. The dynamic model is usually developed on the basis of the static model and uses approximately 70% of its software. In this way, these models constitute a common program complex of the thermodynamic model.

As the thermodynamic model is complex and relatively slow, a number of simplified data-driven models are determined on its basis.

To quantify fault influence in gas turbine diagnostics, a healthy engine performance (baseline model $\vec{Y}_0 = F(\vec{U})$) is required. As mentioned in [8], a simple second-order polynomial function provides accurate approximation of an engine baseline. For one gas path variable Y and three operating conditions u_i, this function looks like

$$Y_0(\vec{U}) = a_0 + a_1u_1 + a_2u_2 + a_3u_3 + a_4u_1u_2 + a_5u_1u_3 + a_6u_2u_3 + a_7u_1^2 + a_8u_2^2 + a_9u_3^2 \tag{3}$$

Unknown coefficients a_j for all the variables $\vec{Y}$ are determined by the least squares method using the data generated by the static nonlinear model as input information. Real data collected under test bed or field conditions can also be used.

Another simplified model obtained in the basis of the thermodynamic model is a linear static model presented by

$$\delta\vec{Y} = H\delta\vec{\Theta} \tag{4}$$

For a fixed operating mode ($\vec{U}$-const), this model relates small relative changes $\delta\vec{Y}$ and $\delta\vec{\Theta}$ of the gas path variables and operating conditions, respectively. The necessary influence coefficients of a matrix H are computed according an expression

$$H_{ij} = \frac{\delta Y_i}{\delta\Theta_j} = \frac{Y_i\left(\vec{\Theta}_j\right) - Y_i\left(\vec{\Theta}_0\right)}{Y_i\left(\vec{\Theta}_0\right)} \Big/ \frac{\Theta_j - \Theta_{0j}}{\Theta_{0j}}. \tag{5}$$

The required values $Y_i\left(\vec{\Theta}_0\right)$ and $Y_i\left(\vec{\Theta}_j\right)$ are calculated by the nonlinear static model which runs one time for a healthy engine state and then one time for each variation $\delta\Theta_j$ introduced by turn in health parameters.

One more model based on the thermodynamic model is called a linear dynamic state space model and is presented by the two following equations:

$$\begin{aligned} \dot{\vec{X}}(t) &= A\overrightarrow{\Delta X}(t) + B\overrightarrow{\Delta U}(t) \\ \overrightarrow{\Delta Y} &= C\overrightarrow{\Delta X}(t) + D\overrightarrow{\Delta U}(t) \end{aligned} \tag{6}$$

where $\vec{X}$ stands for a vector of state space variables and Δ denotes a difference between dynamic and static values of variables. Unknown matrices A, B, C, and D are computed through the static nonlinear model in the same way as the matrix H.

2.2 Thermodynamic model improvements

The development and diagnostic use of thermodynamic models began in the 1970s inspired by the works of Saravanamuttoo (for instance, [9]). Since then, many enhancements have been introduced in this model.

Stamatis et al. [10] proposed an adaptive simulation by an identified thermodynamic model, and then they used such simulation in gas turbine diagnostics [11].

Paper [12] introduces ellipsoid functions that more accurately describe components performance maps in a thermodynamic model. As a result, better model identification at steady states and transients is gained.

In the well-known universal software GasTurb [13] developed since the 1990s, there are special tools that help to verify and correct the compressor and turbine maps.

A more radical way to enhance compressor description is described in [14]. It is proposed to replace a compressor map by a stage-based compressor model. It is shown that the thermodynamic model after such a modification allows to identify faulty stages and recognize compressor fouling, tip wear, and erosion, thus making the diagnosis more profound.

Volponi et al. [15] propose to compliment a traditional thermodynamic model by a neural network-based data-driven model that compensates systematic measurement errors. The authors show that the new hybrid model considerably enhances simulation accuracy.

The above described enhancements contribute to a whole thermodynamic model or its static part. Nevertheless, simulation of transients has specific issues to address, and fast and accurate dynamic models are in demand [16]. Following this demand, paper [17] describes the model of turbine clearance dynamics intended to complement a traditional physics-based dynamic model. Such an extension of the traditional model does not practically change computation time but allows to consider a dynamic turbine performance and significantly enhance the accuracy of engine dynamic simulation.

The present book meets the mentioned demand for improved dynamic models addressing two related problems. The first problem presents the extension of dynamic model operation on engine starting (see Chapter 2). It is proposed to simulate the starting by a linear dynamic model supplemented with a simplified static model. The second problem presents accurate simulation of gas capacities in a nonlinear dynamic physics-based model (Chapter 3). A lot of different variations of capacity models are considered and compared, and the recommendations of application are given.

2.3 Estimation of unmeasured variables

In addition to traditional diagnostic functions, estimation of important unmeasured engine variables can be included into a gas turbine monitoring system.

Examples of such estimation include engine power [18] or thrust [19], compressor and turbine efficiencies [20, 21], and compressor air mass flow [21]. These variables help with monitoring of the mission of gas turbines, their integrity, and overall efficiency. They allow a more profound diagnosis of engine components as well.

The issue of unmeasured variable estimation is challenging because it must allow for an engine technical state. A reasonable solution for an offline monitoring is using a thermodynamic model. However, this model has intrinsic inaccuracy, is critical to computer resources, and is not always available. A Kalman filter-based dynamic model affords a faster estimation [19], but it has additional linearization and approximation errors, and its development still needs a thermodynamic model.

For online monitoring, a good choice is using simple thermodynamic relations that allow computing some important unmeasured variables, for example, component efficiencies [20] and airflows [21], through measured quantities. Nevertheless, this choice is available only for few variables.

In contrast, paper [22] offers a universal data-driven method to estimate any necessary unmeasured gas path variable. Additionally, to draw diagnostic information for online monitoring, it is proposed to extend traditional computing the measured quantity deviations on the unmeasured variables.

Chapter 4 of the present book presents new models of unmeasured variables to monitor engine lifetime. To evaluate the lifetime, the temperatures of gas and air around a critical element, turbine blade, should be known. Various variations of data-driven and physics-based models for these unmeasured variables are formed. By model comparison, the optimal ones are chosen and recommended for real applications.

3. Gas turbine control

The type and configuration of a GT control system is an important factor. They are closely related with the complexity of engine dynamics and control tasks. An improper control system can cause severe damages to engine health. Control programs depend on engine operation modes such as start-up, dynamic operation, steady-state operation, and shutdown. In addition to a specific program for each mode, the control system has various protection programs to avoid overspeed, overheat, flameout, and so on.

Gas turbine control systems are usually of a closed-loop type. The system elaborates a correction to a control variable (e.g., fuel consumption) on the basis of a discrepancy between actual and necessary values of a controlled variable (e.g., rotor speed). To be effective, the closed-loop controller should "know" accurate relations between engine parameters. A conventional approach to controller development relies on generalized engine performances and average operating conditions. However, engine operating and health conditions vary along time.

As advanced gas turbines have several control variables, more flexible control laws can be developed to take into consideration actual engine operating conditions and technical state. Modem digital controllers are capable to realize such complex control and, as a result, minimize fuel consumption, extend engine life, and reduce maintenance costs.

4. Wind turbines

As with gas turbines, wind turbines are an important energy source. However, in contrast to the formers, they have the lowest greenhouse gas emissions and water consumption. Large turbines constitute wind farms to generate electricity for domestic consumption and for the electrical grid. Typically, they have a three-blade horizontal axis construction. The power limit of such a wind turbine is 16/27 times the kinetic energy of the passing air. The power losses include friction of rotor blades, mechanical losses in bearings and gearbox, and losses in a generator and converter. In commercial turbines, these losses can be lessen up to 20–25%. A general trend to increase efficiency and reduce maintenance costs is increasing power and size of a turbine unit. The large wind turbine has a capacity of 9.5 MW, overall height of 220 m, and diameter of 164 m [23].

The efficiency can decrease because of the dust and insects in the air and possible ice accretion resulting in the altered aerodynamic profiles and efficiency losses of 1.2% per year. To monitor turbine performance and structure, accelerometers and strain sensors are usually installed in the nacelle. To assess the dynamics of turbine blades, digital image correlation and stereophotogrammetry are currently applied [24].

As with gas turbines, the design of such large structures as wind turbines needs mathematical modeling. In particular, for developing more accurate control and diagnostic systems, the general-purpose model that describes both aerodynamic and electric parts of the wind power plant is on demand. Chapter 5 of the present book introduces such a model.

Acknowledgements

This work has been carried out with the support of the National Polytechnic Institute of Mexico (research project 20201738).

Author details

Igor Loboda
Instituto Politécnico Nacional, Ciudad de México, México

*Address all correspondence to: igloboda@gmail.com

References

[1] Boyce MP, editor. Gas Turbine Engineering Handbook. 4th ed. Oxford: Elsevier Inc; 2012. 956p. ISBN: 978-0-12-383842-1

[2] Kulikov GG, Thompson HA, editors. Dynamic Modelling of Gas Turbines: Identification, Simulation, Condition Monitoring, and Optimal Control. London: Springer; 2004. 309p. ISBN: 1-85233-784-2

[3] Asgari H, Chen XQ. Gas Turbines Modeling, Simulation and Control Using Artificial Neural Networks. London: CRC Press/Tailor & Francis Group; 2016. 176p. ISBN 13: 978-1-4987-2661-0

[4] Yunusov S, Labendik V, Guseynov S. Monitoring and Diagnostics of Aircraft Gas Turbine Engines. Germany: LAMBERT Academic Publishing GmbH & Co; 2014. 196p. ISBN: 978-3-659-58272-1

[5] Available from: https://ourworldindata.org/energy [Accessed: 01 March 2020]

[6] HIH S, Rogers GFC, Cohen H, editors. Gas Turbine Theory. 5th ed. Edinburg: Person Education Limited; 2001. 491p. ISBN 13: 978-0-13-015847-5

[7] Ogaji SOT, Li YG, Sampath S et al. Gas path fault diagnosis of a turbofan engine from transient data using artificial neural networks. In: Proceedings of IGTI/ASME Turbo Expo 2003, 16–19 June 2003; Atlanta, Georgia, USA: ASME; 2003. p. 10. ASME Paper GT2003-38365

[8] Loboda I. Gas turbine condition monitoring and diagnostics. In: Injeti G, editor. Gas Turbines. Croatia: Sciyo; 2010. pp. 119-144. ISBN: 978-953-307-146-6. Available from: http://www.intechopen.com/articles/show/title/gas-turbine-condition-monitoring-and-diagnostics

[9] Saravanamuttoo HIH, MacIsaac BD. Thermodynamic models for pipeline gas turbine diagnostics. ASME Journal of Engineering for Power. 1983;**105**: 875-884

[10] Stamatis A, Mathioudakis K, Papailiou KD. Adaptive simulation of gas turbine performance. Journal of Engineering for Gas Turbines and Power. 1990;**112**:168-175

[11] Kamboukos P, Mathioudakis K. Multipoint non-linear method for enhanced component and sensor malfunction diagnosis. In: Proceedings of IGTI/ASME Turbo Expo 2006, 8–11 May 2006. USA: ASME; 2006. p. 9. ASME Paper GT2006-90451

[12] Tsoutsanis E, Meskin N, Benammar M, Khorasani K. An efficient component map generation method for prediction of gas turbine performance. In: Proceedings of ASME Turbo Expo 2014, 16–20 June 2014. USA: ASME; 2014. p. 12. ASME Paper GT2014-25753

[13] GasTurb 12. Manual. Germany: GasTurb GmbH; 2015. 294p

[14] Aretakis N, Roumeliotis I, Mathioudakis K. Performance model "zooming" for in-depth component fault diagnosis. Journal of Engineering for Gas Turbines and Power. 2011;**133**(3):1-11

[15] Volponi A, Brotherton T, Luppold R. Empirical tuning of on-board gas turbine engine model for real-time module performance estimation. Journal of Engineering for Gas Turbines and Power. 2008;**130**(3)

[16] Jaw LC, Mattingly JD. Aircraft Engines Controls: Design, System Analysis, and Health Monitoring. Reston, Virginia: American Institute of Aeronautics and Astronautics, Inc; 2009

[17] Yepifanov S, Zelenskyi R, Loboda I. Modeling the gas turbine engine under

its dynamic heating conditions. Journal of Engineering for Gas Turbines and Power - Transactions of ASME. 2015; **137**(3):1-10

[18] Jiang X, Mendoza E, Lin TP. Bayesian calibration for power splitting in single shaft combined cycle plant diagnostics. In: Proceedings of ASME Turbo Expo 2015, 15–19 June 2015. USA: ASME; 2015. p. 11. ASME Paper GT2015-43878

[19] Palmer C, Hettler E. Thrust measurement model-based correction system for turbine engine test cell dynamic data. In: Proceedings of ASME Turbo Expo 2015, 15–19 June 2015. USA: ASME; 2015. p. 8. ASME Paper GT2015-43720

[20] Kacprzynski GJ, Gumina M, Roemer MJ, Caguiat DE, Galie TR, McGroarty JJ. A prognostic modelling approach for predicting recurring maintenance for shipboard propulsion system. In: Proceedings of ASME Turbo Expo 2015, 15-19 June 2015. USA: ASME; 2015 ASME Paper No.2001-GT-0218

[21] Cortés O, Urquiza G, Hernández JA. Optimization of operating conditions for compressor performance by means of neural network inverse. Applied Energy. 2009;**86**:2487-2493

[22] Loboda I, Zárate M, Yepifanov S, Maravilla Herrera C, Ruiz P. Estimation of gas turbine unmeasured variables for an online monitoring system. International Journal of Turbo & Jet Engines. 2018. pp. 1-16. e-ISSN: 2191-0332 (e-publication)

[23] The World's Most Powerful Available Wind Turbine Gets Major Power Boost. UK: MHI Vestas Offshore; 2018. Available from: www.mhivestasoffshore.com [Accessed: 09 March 2020]

[24] Baqersad J, Niezrecki C, Avitabile P. Full-field dynamic strain prediction on a wind turbine using displacements of optical targets measured by stereophotogrammetry. Mechanical Systems and Signal Processing. 2015; **62–63**:284-295. DOI: 10.1016/j.ymssp.2015.03.021

Section 2

Gas Turbines

Chapter 2

Turbine Engine Starting Simulation

Sergiy Yepifanov and Feliks Sirenko

Abstract

The process of engine control development requires the models that describe engine operation and its response on a control action. The development flow required numerous models to be engaged, like component-level non-linear model, engine-level non-linear model, linear dynamic model, etc. Models made a great progress during the recent years and became reliable tools for control engineers. However, most models are derivatives from the component-level non-linear model, which in its turn consumes the component performances. Things turn different when one addresses the starting range of engine operation. The problem here is all about the missing performances of the engine components, as it is quite hard to harvest these performances in this region as the processes that happen in the engine are transient by nature. Different scientists offered different approaches to the problem of building the component level non-linear model of the sub-idle region, but the general idea is to somehow extrapolate the known performances to the sub-idle region. However, there are no known reports about a model that considers all aspects of this approach and simulates the engine starting. In this chapter, you can find an alternative view on a problem of simulation of a sub-idle operation. The proposed model belongs to a group of linear dynamic models including the static model as well as simplified static model to support the dynamic model. Instead of trying to extrapolate component performances and get the full-scale component-level model, you will see that the canonical component performances are replaced by the direct relations between parameters that are used in the control algorithms, like gas-path parameters against the RPM. As well in this chapter, you will find the exact instructions on how to create the model and an example of the one with the real test data.

Keywords: turbine engine, starting, mathematical model, experimental data, identification, approximation

1. Introduction

Models are known for their benefit to reduce the price for engine development replacing the numerous experiments with the simulations. This has the biggest value for the control engineers as they need the greatest amount of experimental data. Industry came to a gold standard on how to simulate the models of engine operation within the range limited with idle on one side and the maximum on the other side. These models are well structured and validated, having the well-known thermodynamic relations under the hood. But their usage is impossible or considerably limited in the sub-idle range, because of no experimentally registered performances of engine components. The other reason is that the required performances are challenging to be determined for the sub-idle region, especially the performance of a combustion chamber.

Here we have reached the point that suites the most to introduce the existing models of sub-idle operation. They can go into two baskets. In the first we find the methods that deal with the extrapolation of the component performances into the sub-idle range, while in the second, methods that approximate the experimental data with various polynomials.

We can dive deeper to the first basket and discriminate four groups of models

- based on straight extrapolation of known performances (no physics considered) [1–3];
- "smart" extrapolation that consumes the theoretical knowledge about component performances in the sub-idle region [4–7];
- statistical extrapolation [8];
- based on fitting the generic performances to the experimental data [9–11].

The brightest method of the first group is based on the simplest similarity laws. The method was developed by Gaudet et al. [1]. The method is only applicable for incompressible fluid. Another method by Sexton [2] suggests obtaining new constant corrected speed lines of compressor spool using experimentally measured ones:

$$\frac{(W_a)_{i-1}}{(W_a)_i} = \left[\frac{N_{i-1}}{N_i}\right], \frac{(L_{sp})_{i-1}}{(L_{sp})_i} = \left[\frac{N_{i-1}}{N_i}\right]^2, \ (Pow)_{i-1}/(Pow)_i = [N_{i-1}/N_i]^3. \quad (1)$$

where subscript i refers to experimentally measured parameter and $i-1$ refers to extrapolated parameter.

Gaudet and Gauthier improved Sexton's method by adopting it for compressible fluid. Powers in Eq. (2) are obtained by the analysis of two experimentally measured corrected constant speed lines with the lowest values of corrected rotational speed. Next, a new corrected constant speed line is calculated referring existing by the equations

$$\frac{(W_a)_{i-1}}{(W_a)_i} = \left[\frac{N_{i-1}}{N_i}\right]^p, \frac{(L_{sp})_{i-1}}{(L_{sp})_i} = \left[\frac{N_{i-1}}{N_i}\right]^q; (Pow)_{i-1}/(Pow)_i = \left[\frac{N_{i-1}}{N_i}\right]^r. \quad (2)$$

Powers p, q, and r stay constant for all sub-idle operating range.

The second group is very similar to the first one and differs in the fact that extrapolation takes into account prior information about processes taking place during starting. This allows improving the tolerance and preventing errors.

All methods of the second group consider only separate facts from prior information (e.g. negative efficiency of compressor spool till the moment of combustion chamber lighting [5]), but there are still no known methods that are able to take into account all prior information. For example, there is a method [6, 7] to extrapolate performances of compressor or turbine taking into account the continuity equation violation at negative efficiency. Authors suggest routine performance representation to be changed to the new one. The efficiency performance of compressor spool is expressed as follows:

$$\eta_C^* \cdot \left[L_C^*/N^2\right] = f\left(W_a \cdot T_{in}^*/N \cdot P_{in}^*\right). \quad (3)$$

Another method, proposed by Agrawal and Yunis [8], belongs to methods of the third group. This method requires a big number of experimental data for identifying true performances of engine components from universal ones. Relations to carry out identification are the following:

- Air flow

$$\frac{\left(W_a \cdot \sqrt{T_{cor}}/P_{cor}\right)_{i-1}}{\left(W_a \cdot \sqrt{T_{cor}}/P_{cor}\right)_i} = K_\phi \cdot \frac{\left(N/\sqrt{T_{cor}}\right)_{i-1}}{\left(N/\sqrt{T_{cor}}\right)_i}; \quad (4)$$

- Specific work

$$\frac{\left(L_{sp}/T_{cor}\right)_{i-1}}{\left(L_{sp}/T_{cor}\right)_i} = K_\psi \cdot \left[\frac{\left(N/\sqrt{T_{cor}}\right)_{i-1}}{\left(N/\sqrt{T_{cor}}\right)_i}\right]^2; \quad (5)$$

- Efficiency

$$\frac{(\eta_C)_{i-1}}{(\eta_C)_i} = K \cdot \eta_{C\,i} \cdot \left[\frac{\left(N/\sqrt{T_{cor}}\right)_{i-1}}{\left(N/\sqrt{T_{cor}}\right)_i}\right], \quad (6)$$

where K_ϕ, K_ψ, and K are coefficients depending on a corrected rotational speed. This relation is formed from analysis of a big number of similar engines.

In addition to the above parameters, this method also allows simulation of torque, pressure, and temperature at compressor discharge.

The last group of methods can be represented by Kong's method [9]. The point of the method is getting universal component performances from well-known thermodynamic relations for gas turbine engines and prior information about the engine with its further identification at modes from idle to maximum power (thrust). Here, identification must be understood as scaling factors calculating to match theoretical and true performances. This method has validated its effectiveness for the set of turboshaft engines. But the shortage of this method was poor convergence at modes with low rotational speeds, including sub-idle modes.

To cope with the problem of poor convergence, Jones et al. had improved Kong's method, suggesting application of non metering components and new criterion for identification quality estimation (surge parameter). Applied measures have improved identification, especially at sub-idle modes [12].

Final improvement of Kong's method was carried out by the group of scientists from Sharif University. They have suggested a new approach to scaling factor calculation. This approach provides good identification quality at all modes. All parameters necessary for compressor map identification can be calculated as follows [13]:

$$W_{a\,cor} = \left(W_{a\,cor}\right)_{univ} \cdot W_{a\,cor\,calc}/\left(W_{a\,cor\,calc}\right)_{univ}, \quad (7)$$

$$N_{cor} = \left(N_{cor}\right)_{univ} \cdot N_{cor\,calc}/\left(N_{cor\,calc}\right)_{univ}, \quad (8)$$

$$\pi = \pi_{univ} \cdot (\pi_{calc} - 1) + 1/(\pi_{calc\,univ} - 1) + 1, \quad (9)$$

$$\eta = \eta_{univ} \cdot \eta_{\text{calc}} / \eta_{\text{calc univ}}. \tag{10}$$

Methods of the second category accept the hypothesis about similarity of starting processes in all engines. Here appears the conclusion that any starting process can be circumscribed by universal relations between control factors and measured parameters. These methods require storing rig and in-flight experimental information to form appropriate data for identification. The polynomial relations are conserved to have the same structure from engine to engine. Only coefficients of polynomial change [14].

Wrapping up the above overview of the methods to simulate sub-idle operation of the gas turbine, one can conclude the following

- most described methods can hardly find the usage in an every day's engineering practice;
- most methods address the compressor map extrapolation ignoring the problem of other compressor components, especially the combustion chamber operation;
- to identify the performances from the experimental data, the latest must be of high quality, however, as the starting is a dynamic process, the collected data may be corrupted with time lag as well as poor sensor performance at low operational modes;
- getting the accurate experimental data to build the high-quality performance map is not always a good value for money as it usually needs mounting extra sensors and carrying out additional experiments (for example, low-inertia thermocouple, strain gage, etc.);
- the methods from the second basket have extremely limited application range as they require the same starter to be used as well as the minor changes to the starting control loop are allowed.

2. Requirements to the starting simulation model

During the starting engine speeds up from either turned off state to an idle running or windmilling to the flight idle. Based on the source of power that drives the rotor during the acceleration, the whole starting may be decomposed into three phases.

The rotor of the engine is dynamically balanced. During the first phase, the rotor is driven by the starter only, as the combustion chamber is not lighted yet. During this phase, the starter drives the rotor from turned off state to the RPM where the combustion chamber is started. The torque balance of this phase is

$$M_{starter} = M_{\text{res}} + M_{inert}, \tag{11}$$

where M_{res} is the total torque of resistance that is generated by compressor and turbine cascades of the rotor, by bearings and driven auxiliary units (fuel and oil pumps, electric generator, deaerator, etc.).

As soon as combustion chamber is brought into a game, the turbine starts giving the positive input to the torque balance equation. Both turbine and starter drive the rotor because the turbine does not have enough power yet to accelerate the rotor alone. The second phase lasts till the turbine reaches the power where it can drive the compressor and accelerate the rotor on its own. The torque equation for the second phase is

$$M_{starter} + M_T = M_{res} + M_{inert}. \tag{12}$$

The starting ends up with the phase where the starter is moved from the table, making the turbine to be the only one source of power that drives the compressor and makes all rotors accelerate to the idle or the flight idle (see **Figure 1**). The torque balance equation is

$$M_T = M_{res} + M_{inert}. \tag{13}$$

The first and foremost problem that every simulation engineer faces when designing the model is the problem of the choice among the available model structures, as the selected structure must make a perfect fit to the problem to be solved by the model. The major goal of the proposed model is to perform the preliminary tuning of the ACS. For this, the model must:

- provide correct relations among parameters that describe the operation of the engine (as the starting is a non-steady-state process, the model must properly describe the dynamic properties of GTE including all processes that determine engine dynamics);
- be able to be integrated with the vast majority of the models that are normally used to support the ACS design, e.g. model that describes the engine operation in the range above the idle, models of governors, etc.;
- properly describe the features typical for the starting, for example, parameters overlap during the starting;
- require minimum amount of experimental data to be generated and adjusted;
- give a quantitative assessment to the required parameters that cannot be measured directly.

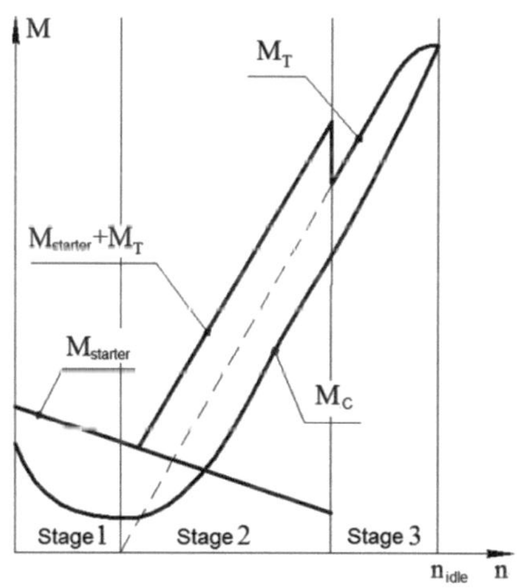

Figure 1.
Engine starting stages and moment performances.

3. Structure

Form the vast majority of model structures, the good choice to fulfill the requirements described above is the dynamic model consisting of linear dynamic model,

$$\begin{cases} \dot{\vec{X}} = A \cdot \Delta\vec{X} + B \cdot \Delta\vec{U}; \\ \Delta\vec{Y} = C \cdot \Delta\vec{Y} + D \cdot \Delta\vec{U}. \end{cases} \tag{14}$$

supported with steady-state model in simplified representation (next—simplified static model, SSM).

The structure of the model is presented in a **Figure 2**.

The model of the three-spool turbofan must consider the mutual interaction of engine spools.

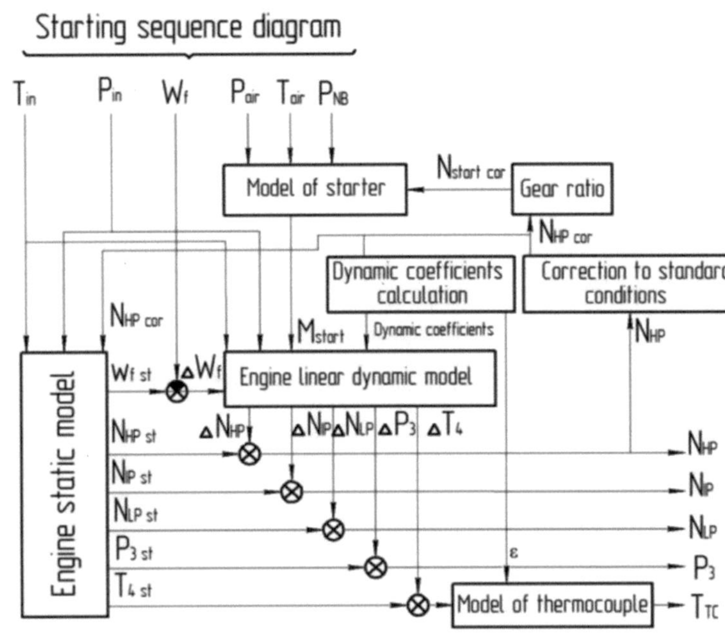

Figure 2.
Structure of the proposed starting model of a turbofan.

4. Compiling of the SSM

Static performance of GTE in a sub-idle region is nothing more than an abstraction, as in a sub-idle region, the engine does not have any steady modes. The SMM improves the precision of the starting model as well as provides the unique architecture of both sub-idle and above idle models. Let the SSM be formalized as $\vec{Z} = f(\overline{N})$, where $\vec{Z} = \left[\vec{X}^T \vec{Y}^T\right]^T$; $\overline{N} = \frac{N_{HP\ cor}}{N_{HP\ 0}}$ is the base value that corresponds to an idle mode.

The proposed scheme of SMM for one of parameters Y from vector in the sub-idle region is presented in **Figure 3**.

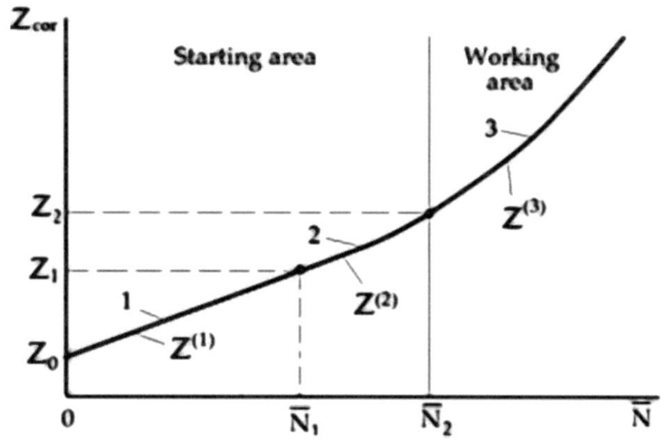

Figure 3.
Static model generation: 1—linear segment; 2—parabola; 3—working line.

The static performance is extrapolated to the sub-idle region with a linear region and parabolic region (linking region). The performance is described with the functions. They are limited on both sides of the region as

- values of the functions must be equal on the left and on the right;
- first derivative of the functions must be equal on the left and on the right.

Such a structure allows alternating the slope of the linear region within some range (see line 1 in **Figure 3**). This is a good opportunity to adjust the SSM after the experiments have been carried out.

Equation of linear region is described as

$$Z^{(1)} = Z_0 + k\overline{N}. \tag{15}$$

Initial values of slope angle are evaluated from the next formula

$$k = (0.6 \ldots 1.2)\frac{Z_2 - Z_0}{\overline{N}_2}. \tag{16}$$

The 0.6–1.2 range determines the adjustment range of the linear region. If k = 1, then linear region matches with the line joining two points $(0, Z_0)$, $(\overline{N}_2, Z_2)$. Y_0 is equal to a default value of a parameter to be simulated (temperature, pressure, RPM, etc.).

The linking function is described as

$$Z^{(2)} = k_0 + k_1\overline{N} + k_2\overline{N}^2. \tag{17}$$

Unknown coefficients k_0, k_1, and k_2 are evaluated from the set of linear equations that is compiled from the limitations for the regions, described in this chapter:

$$Z_0 + k\overline{N}_1 = k_0 + k_1\overline{N}_1 + k_2\overline{N}_1^2; \tag{18}$$

$$k = k_1 + 2k_2\overline{N}_1; \tag{19}$$

$$k_0 + k_1\overline{N}_2 + k_2\overline{N}_2^2 = Z_2; \tag{20}$$

$$k_1 + 2k_2\overline{N}_2 = \left(\frac{dZ}{d\overline{N}}\right)_2. \tag{21}$$

The last unknown to be determined from the equation set (18)–(21) is an abscissa $\overline{N}_1$ of point joining linear function with the linking function.

Set of Eqs. (18)–(21) is non-linear. Let us use Eqs. (20, 21) to solve it and express the coefficients k_0 and k_1 to be dependent on k_2:

$$k_1 = \left(\frac{dZ}{d\overline{N}}\right)_2 - 2k_2\overline{N}_2;\ k_0 = Z_2 - \left(\frac{dZ}{d\overline{N}}\right)_2 \overline{N}_2 + k_2\overline{N}_2^2. \tag{22}$$

From Eq. (19), we get the relation between $\overline{N}_1$ and k_2:

$$\overline{N}_1 = \frac{k - D + 2k_2\overline{N}_2}{2k_2}. \tag{23}$$

Putting the obtained equations into (18), we get the solution:

$$k_2 = \frac{\left(k - \left(\frac{dZ}{d\overline{N}}\right)_2\right)^2}{4(Z_2 - Z_0 - k_2\overline{N}_2)};\ k_1 = \left(\frac{dZ}{d\overline{N}}\right)_2 - \frac{\left(k - \left(\frac{dZ}{d\overline{N}}\right)_2\right)^2\overline{N}_2}{2(Z_2 - Z_0 - k_2\overline{N}_2)}; \tag{24}$$

$$k_0 = Z_2 - \left(\frac{dZ}{d\overline{N}}\right)_2 \overline{N}_2 + \frac{\left(k - \left(\frac{dZ}{d\overline{N}}\right)_2\right)^2\overline{N}_2^2}{4(Z_2 - Z_0 - k_2\overline{N}_2)};\ \overline{N}_1 = \frac{2(Z_2 - Z_0 - k_2\overline{N}_2)}{\left(k - \left(\frac{dZ}{d\overline{N}}\right)_2\right)} + \overline{N}_2. \tag{25}$$

Building up any performance, we will require turned-off parameters, idle parameters and derivative of the performance parameter against the mode parameter, e.g. RPM of HPR. An example of this map with experimental points is shown in **Figure 4**.

Because the corrected exhaust gas temperature $T_{T\ cor}$ has another physical nature, its relation is formed using another approach method.

The turbine temperature decreases to some value when the mode decreases. Further mode decrease results in a drastic temperature increase mainly because of the inefficient turbine operation and low airflow through the engine.

To consider this prior information, the turbine discharge temperature is modeled as follows:

$$T_{4\ cor} = T_{2\ cor} + K_T \frac{W_{f\ cor} \cdot H_u}{W_{air\ cor} \cdot c_p}, \tag{26}$$

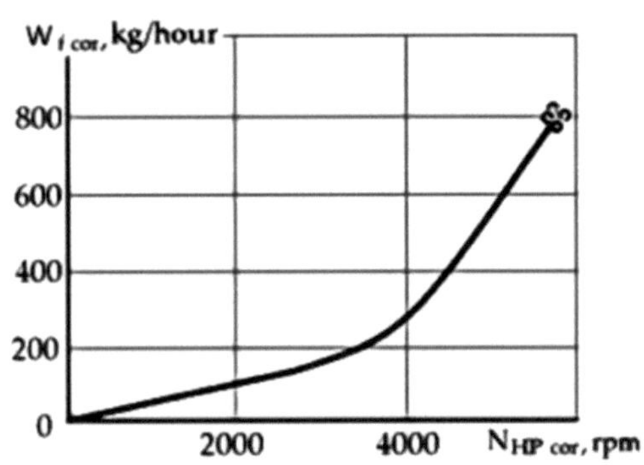

Figure 4.
Relation $W_{f\ st\ cor} = f(N_{HP\ cor})$ of the SSM.

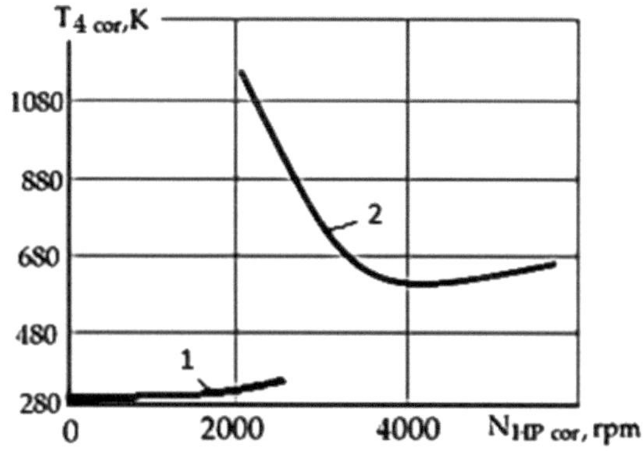

Figure 5.
Static model of the gas temperature.

where $T_{2\ cor}$ is the temperature of airflow at compressor discharge; $\frac{W_{f\ cor} \cdot H_u}{W_{air\ cor} \cdot c_p}$ is the temperature rise caused by the amount of heat added by the fuel combustion in the combustion chamber.

A coefficient K_T considers the temperature drop that is caused by heat retraction in a high-pressure turbine and a free turbine. The coefficient is calculated to provide the equality of the exhaust gas temperatures. The first and the second values of the exhaust-gas temperature are calculated using the sub-idle operation model and the above-idle operation model, respectively.

The airflow through the engine is a priori known to be proportional to the square of the rotational speed: $W_{air\ cor} = C \cdot N_{HP}^2$.

The dependence $T_{4\ cor} = f(N_{HP\ cor})$ is shown in **Figure 5**. The left branch (1) corresponds to the Stage 1 of starting (see **Figure 1**) when only starter rotates the rotor; the right branch presents Stages 2 and 3 when the combustion chamber is switched in.

It follows from Eq. (10) that for proper modeling of gas temperature on must compile the SSMs of $T_{2\ cor}$ and airflow $W_{air\ cor}$.

5. Identification of a linear dynamic model

The above dynamic model (1) in the state space for the three-spool turbofan engine under consideration has the form of a system of linear algebraic and differential equations.

$$\begin{cases} \dot{N}_{IP} = a_{11} \cdot \Delta N_{IP} + a_{12} \cdot \Delta N_{HP} + a_{32} \cdot \Delta N_{LP} + b_{11} \cdot \Delta W_f; \\ \dot{N}_{HP} = a_{21} \cdot \Delta N_{IP} + a_{22} \cdot \Delta N_{HP} + a_{23} \cdot \Delta N_{LP} + b_{21} \cdot \Delta W_f + b_{24} \cdot (M_{start} - M_{res}); \\ \dot{N}_{LP} = a_{31} \cdot \Delta N_{IP} + a_{32} \cdot \Delta N_{HP} + a_{33} \cdot \Delta N_{LP} + b_{31} \cdot \Delta W_f; \\ \Delta p_2 = c_{11} \cdot \Delta N_{IP} + c_{12} \cdot \Delta N_{HP} + c_{13} \cdot \Delta N_{LP} + d_{11} \cdot \Delta W_f; \\ \Delta T_4 = c_{21} \cdot \Delta N_{IP} + c_{22} \cdot \Delta N_{HP} + c_{23} \cdot \Delta N_{LP} + d_{21} \cdot \Delta W_f, \end{cases} \tag{27}$$

where $\Delta N_{IP} = N_{IP} - N_{IP\ st}$, $\Delta N_{HP} = N_{HP} - N_{HP\ st}$, $\Delta N_{LP} = N_{LP} - N_{LP\ st}$, $\Delta W_f = W_f - W_{f\ st}$, $\Delta p_2 = p_2 - p_{2\ st}$ and $\Delta T_4 = T_4 - T_{4\ st}$.

This system can be resolved by the Euler method.

The set of Eqs. (27) can be simplified by the introduction of the listed next assumptions:

- as the argument of the static model and functions for determining the dependence of the coefficients of the linear dynamic model on the engine operational mode is N2, then $N_{HP} = N_{HP\ st}$ which is why $\Delta N_{HP} = 0$.

- the mutual effect of the rotors is determined mainly by the interaction of the spools of the turbine; therefore, it spreads along the flow (the high-pressure rotor affects the intermediate and low-pressure rotor, and the intermediate-pressure rotor affects the low-pressure rotor).

Concluding the just listed assumptions, the set of Eqs. (27) transforms to

$$\begin{cases} \dot{N}_{IP} = a_{11} \cdot \Delta N_{IP} + b_{11} \cdot \Delta W_f; \\ \dot{N}_{HP} = b_{21} \cdot \Delta W_f + b_{24} \cdot (M_{start} - M_{res}); \\ \dot{N}_{LP} = a_{31} \cdot \Delta N_{IP} + a_{33} \cdot \Delta N_{HP} + b_{31} \cdot \Delta W_f; \\ \Delta p_2 = d_{11} \cdot \Delta W_f; \\ \Delta T_4 = d_{21} \cdot \Delta W_f. \end{cases} \tag{28}$$

The task of synthesizing a linear dynamic model is reduced to the problem of determining the coefficients of a linear dynamic model.

The coefficients b_{24}, b_{21}, and d_{11} are determined analytically using a priori and experimental information, followed by approximation by polynomial dependencies, the arguments of which are the rotational speed of the high-pressure rotor. The coefficient b_{24} was determined analytically from the following equation:

$$J_{HP} \frac{d\omega_{HP}}{dt} = M_{start} - M_{HPC}\left(N_{HP}^2\right), \tag{29}$$

where J_{HP} is an inertia of high pressure rotor.

Having transformed $\frac{d\omega_{HP}}{dt}$ to $\frac{\pi}{30} \cdot \frac{dN_{HP}}{dt}$, we get:

$$b_{24} = \frac{30}{\pi \cdot J_{HP}} \tag{30}$$

The coefficient b_{24} does not depend on the type or characteristics of the used starter and is constant for the entire starting.

The determination of the coefficient $b_{21\ пр}$ showing the effect of fuel consumption on the rate of the rotational speed change for the high-pressure rotor was carried out analytically according to the following relationship:

$$b_{21\ cor} = \frac{\dot{N}_{HP} - b_{24} \cdot M_{start}}{\Delta W_f}. \tag{31}$$

Further, the obtained values were approximated, considering the following a-priori information about this coefficient at the starting region. Let us consider the following equation:

$$\tau_{HP} \frac{dN_{HP}}{dt} + \Delta N_{HP} = K_W \cdot \Delta W_f, \tag{32}$$

where τ_{HP} is the time constant of the high-pressure rotor.

Let us divide both parts of this equation by time constant and move ($\frac{1}{\tau_{HP}} \Delta N_{HP}$) to the right part of the equation, and we get

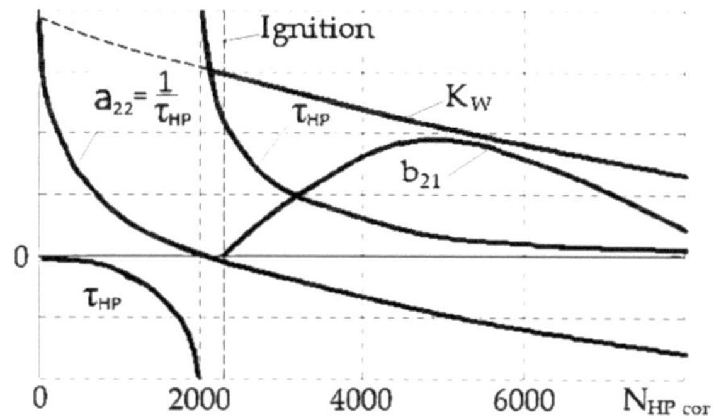

Figure 6.
Coefficients of the high-pressure rotor dynamics.

$$\dot{N}_{HP} = -\frac{1}{\tau_{HP}}\Delta N_{HP} + \frac{K_W}{\tau_{HP}} \cdot \Delta W_f = a_{22}\Delta N_{HP} + b_{21}\Delta W_f. \quad (33)$$

It is known that the time constant at the beginning of the rotor spinning ($N_{HP} = 0$) is equal to zero. At low rotational speeds, the rotor is unstable; therefore, the value of the time constant is negative. Physically, this can be explained by the fact that the turbine in this mode of operation is inefficient, and the slope of the torque characteristic of the turbine (the dependence of the moment on rotation speed at constant fuel consumption) is greater than the slope of the torque characteristic of the compressor. As rotor speed increases, the time constant decreases until the moment when the slope of the torque characteristic of the compressor becomes equal to the slope of the torque characteristic of the turbine $\left(\frac{\partial M_{HPC}}{\partial n_{HP}} = \frac{\partial M_{HPT}}{\partial n_{HP}}\right)$. At this point, the function of the dependence of the time constant on the rotor speed experiences a gap $\left(\tau_{HP} = \frac{\frac{\pi}{30} \cdot J_{HP}}{\frac{\partial M_{HPC}}{\partial N_{HP}} - \frac{\partial M_{HPT}}{\partial N_{HP}}} = \infty\right)$. Thus, at low rotational speeds, the rotor cannot be stable without an additional energy source or closed-loop control system. With a further increase in the rotational speed, the time constant decreases from infinity to the value at the idle mode.

The coefficient K_W has physical sense, therefore, the nature of its change is known. The ratio of this coefficient to the time constant shows the effect of changes in fuel consumption on the rate of N_{HP} change. The dependences of the LDM coefficients of the high-pressure rotor on its rotational speed are shown in **Figure 6**.

The analytical determination of the coefficients a_{11}, b_{11}, a_{31}, a_{33}, and b_{31} is impossible; therefore, they are determined by the least squares method engaging the theoretical and experimental information and with the subsequent approximation of the dependence of the obtained values on a parameter that determines the engine operating mode.

So, for example, the evaluation of coefficients a_{11} and b_{11} is.

$$\hat{\vec{\theta}} = \arg\min J\left(\vec{\theta}\right); \quad (34)$$

where $\vec{\theta} = \begin{bmatrix} a_{11} \\ b_{11} \end{bmatrix}$ is a vector of coefficients to be identified (a_{11}, b_{11}), the change of which is approximated with polynomials $b_{11\,cor} = \sum_{i=0}^{2} q_i \cdot N^i_{HP\,cor}$ and $a_{11\,cor} = \sum_{i=0}^{2} p_i \cdot N^i_{HP\,cor}$, $\hat{\vec{\theta}}$ —vector estimate $\vec{\theta}$.

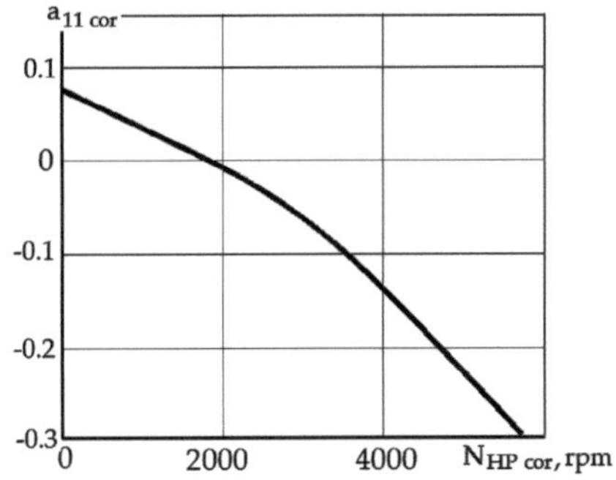

Figure 7.
Relation of coefficient a_{11} on operation mode.

Thus, the problem of the evaluation of coefficients a_{11}, b_{11} reduces to the problem of determining the coefficients of polynomials approximating the change of these coefficients in the sub-idle region:

$$J\left(\vec{\theta}\right) = J(q_2, q_1, q_0, p_2, p_1, p_0) = \sum_{j=1}^{N} \left[\dot{N}_{IP\ cor\ j} - \left\{\left(q_2 N^2_{HP\ cor\ j} + q_1 N_{HP\ cor\ j} + q_0\right)\Delta W_{f\ cor\ j} + \left(p_2 N^2_{HP\ cor\ j} + p_1 N_{HP\ cor\ j} + p_0\right)\Delta N_{IP\ cor\ j}\right\}\right]^2. \tag{35}$$

As an example, **Figure 7** represents the results of a_{11} approximation on the engine operational mode. They correspond to the obtained dependence

$$a_{11\ cor} = -4.0276 \cdot 10^{-9} N^2_{HP\ cor\ j} - 3.0863 \cdot 10^{-5} N_{HP\ cor\ j} + 0.0716. \tag{36}$$

It is known that the coefficient smoothly increases when the rotational speed goes down; therefore, the results of its identification are reliable.

For a more adequate model, the coefficient is refined according to the following relationship:

$$b_{11} = \frac{\dot{N}_{IP\ exp} - a_{11} \cdot \left(N_{IP\ exp} - N_{IP\ st}\right)}{W_{f\ exp} - W_{f\ st}}. \tag{37}$$

6. Model of thermocouple

Measuring the temperature at the sub-idle modes is challenging because the thermocouple has an extremely high time constant. Estimating the real temperature requires complementing the linear dynamic model with an additional equation

$$\varepsilon \frac{dT_{TC}}{dt} + T_{TC} = T_4, \tag{38}$$

where ε is the time constant of the thermocouple.

The time constant experimental observation requires mounting two thermocouples at the same station: a regular thermocouple and a quick-response thermocouple. Then, the stored data are analyzed to find the temperature delay in time.

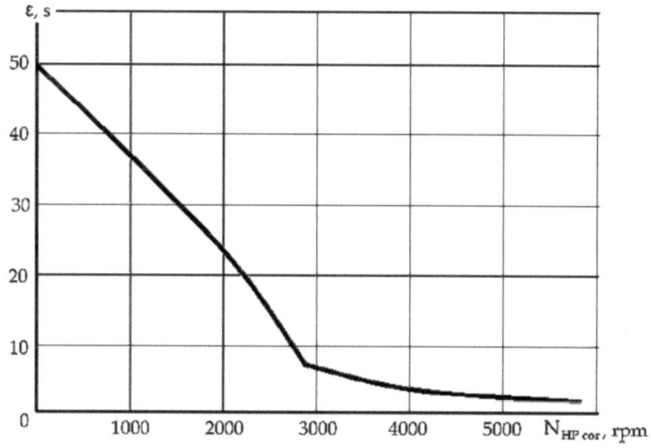

Figure 8.
Estimated time constant of thermocouple as a function of rotation speed.

Unfortunately, the performance of this experiment is notably rare. Hence, the time constant is generally unknown.

Fortunately, this constant can be estimated using the one-time special analysis of the experimental data. The time constant can be identified for the sub-idle modes by substituting T_T in Eq. (38) using the last equation of the system (27):

$$\varepsilon = \frac{c_{21} \cdot \Delta N_{IP} + c_{22} \cdot \Delta N_{HP} + c_{23} \cdot \Delta N_{LP} + d_{21} \cdot \Delta W_f - T_{TC}}{\frac{dT_{TC}}{dt}}, \quad (39)$$

where $\frac{dT_{TC}}{dt}$ and T_{TC} are determined from experimental data.

The resultant time constant of thermocouple as a function of corrected rotation speed is shown in **Figure 8**.

7. Starting simulation algorithm

Starting simulation algorithm implements the structure, which is presented in **Figure 1**, as the following sequence of stages:

1. Initial values of parameters are set as $N_{HP\ cor} = N_{IP\ cor} = N_{LP\ cor} = 0$, $W_f = 0$, $P_2 = P_{in}$, $T_4 = T_2 = T_{TC} = T_{in}$, $W_{air\ cor} = 0$.

2. Knowing $N_{HP\ cor}$, P_{in}, T_{in} (in first step, the initial values are used) and applying the static model ($W_{f\ st\ cor} = f(N_{HP\ cor})$, $N_{IP\ st\ cor} = f(N_{HP\ cor})$, $N_{LP\ st\ cor} = f(N_{HP\ cor})$, $P_{2\ st\ cor} = f(N_{HP\ cor})$, $T_{2\ st\ cor} = f(N_{HP\ cor})$, $W_{air\ st\ cor} = f(N_{HP\ cor})$, $T_{4\ st\ cor} = f(N_{HP\ cor})$), static values of the simulated parameters are calculated: $W_{f\ st\ cor}$, $N_{IP\ st\ cor}$, $N_{LP\ st\ cor}$, $P_{2\ st\ cor}$, $T_{2\ st\ cor}$, $W_{air\ st\ cor}$, $T_{4\ st\ cor}$.

3. Physical values of the parameters $W_{f\ st}$, $N_{IP\ st}$, $N_{LP\ st}$, $P_{2\ st}$, $T_{2\ st}$, $W_{air\ st}$, $T_{4\ st}$ are calculated using equations of correction.

4. Knowing $N_{HP\ cor}$, P_{in}, T_{in} and using the relations $a_{11\ cor} = f(N_{HP\ cor})$, $b_{11\ cor} = f(N_{HP\ cor})$, $b_{21\ cor} = f(N_{HP\ cor})$, $a_{31\ cor} = f(N_{HP\ cor})$, $a_{33\ cor} = f(N_{HP\ cor})$, $b_{31\ cor} = f(N_{HP\ cor})$, $d_{11\ cor} = f(N_{HP\ cor})$, $d_{21\ cor} = f(N_{HP\ cor})$, the coefficients of the linear

dynamic model are determined for a current step of the solution procedure: $a_{11\ cor}$, $b_{11\ cor}$, $b_{21\ cor}$, $a_{31\ cor}$, $a_{33\ cor}$, $b_{31\ cor}$, $d_{11\ cor}$, $d_{21\ cor}$.

5. Using the equations of correction, physical values of the LM coefficients are determined: a_{11}, b_{11}, b_{21}, a_{31}, a_{33}, b_{31}, d_{11}, d_{21}.

6. Deviations ΔU, $\Delta\vec{X}$ (see Eq. (27)) for the current step of the solution procedure are calculated:

$$\Delta U = \Delta W_f = W_f\text{-}W_{f\ st};$$
$$\Delta x_1 = \Delta N_{IP} = N_{IP}\text{-}N_{IP\ st};\ \Delta x_2 = \Delta N_{HP} = 0; \Delta x_3 = \Delta N_{LP} = N_{LP}\text{-}N_{LP\ st}.$$

7. The starter torsion torque M_{start} is determined. The model of starter (see **Figure 2**) contains characteristics of the starter, which determines the starter torsion torque as a function of the starter rotation speed N_{start}, pressure and temperature of the air at the starter inlet P_{air}, T_{air}, and the air pressure at the starter nozzle vanes P_{NB}. The moment of the starter switching off is determined according to the starting sequence diagram. After this moment, M_{start} is set to be zero.

8. Using the quasi-linear dynamic model, new values of the simulated parameters N_{HP}, N_{IP}, N_{LP}, P_2, T_4, T_{TC} are calculated:

$$\begin{cases} N_{IP\ i+1} = N_{IP\ i} + t_{step} \cdot (a_{11} \cdot \Delta N_{IP} + b_{11} \cdot \Delta W_f); \\ N_{HP\ i+1} = N_{HP\ i} + t_{step} \cdot (b_{21} \cdot \Delta W_f + b_{24} \cdot M_{start}); \\ N_{LP\ i+1} = N_{LP\ i} + t_{step} \cdot (a_{31} \cdot \Delta N_{LP} + a_{33} \cdot \Delta N_{LP} + b_{31} \cdot \Delta W_f); \\ P_{2\ i+1} = P_{2\ st\ \ i+1} + d_{11} \cdot \Delta W_f; \\ T_{4\ i+1} = T_{4\ st\ i+1} + d_{21} \cdot \Delta W_f; \\ T_{TC\ i+1} = T_{TC\ i+1} + \dfrac{t_{step}}{\varepsilon}(T_{4\ i+1}\text{-}T_{TC\ i}). \end{cases} \tag{40}$$

9. $N_{HP\ cor}$ is determined on known N_{HP}.

Points 2–9 of this algorithm are repeated up to the end of the starting sequence diagram.

8. Starting model verification

For adequacy of the model and its quality checking, the simulation results were compared with experimental data. For this purpose, the starting process was chosen, which was not used for the model identification and differs from other experiments by a law of the fuel supplying and by ambient conditions. The experimental fuel flow (see **Figure 9a**) was inputted to the model. The time moments of the combustion ignition and the starter switching off are the same as at experiments, which were used for the model identification. The results are shown in **Figure 9b–f**.

Comparison of the simulation results with experimental data shows that the model meets all demands to models, which are used for the ACS development. Visible difference between experimental and simulated data is explained by the

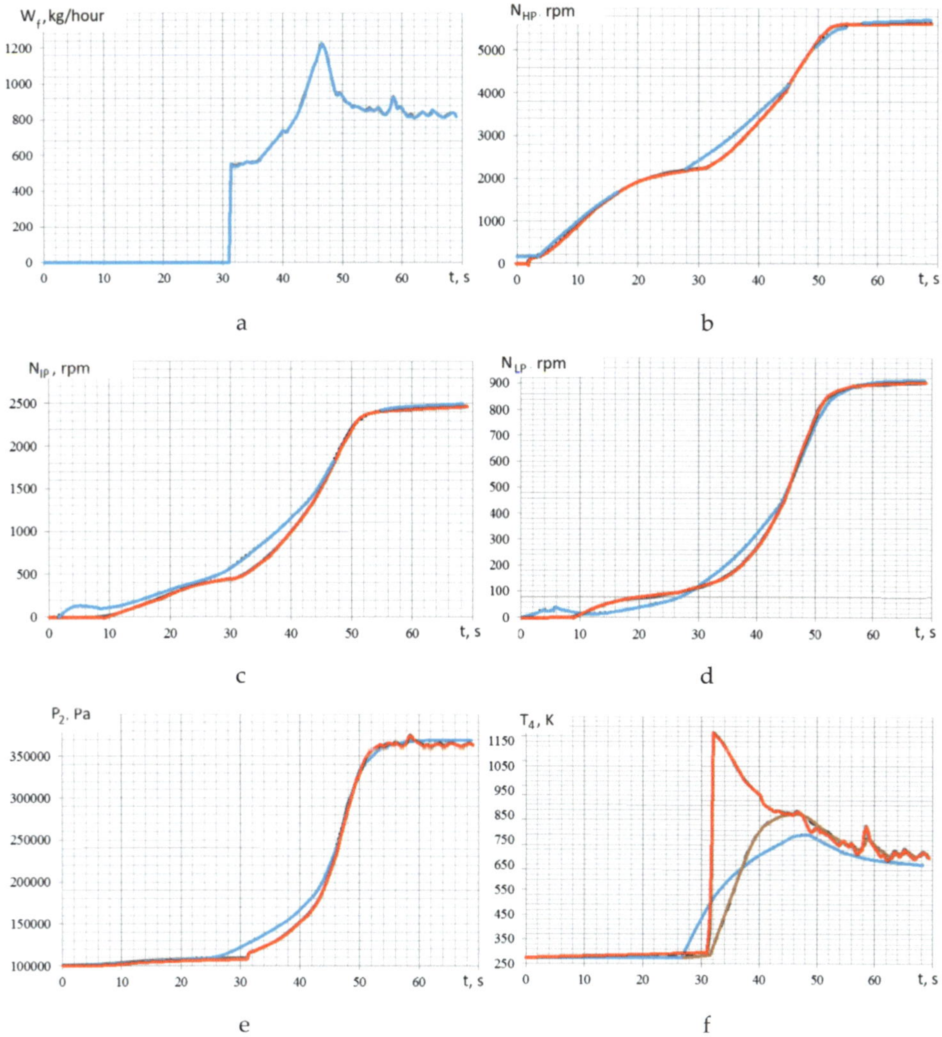

a – experimental fuel flow; b – N_{HP} ; c – N_{IP} ; d – N_{LP} ; e – P_2 ; f – T_4 and T_{TC} ;

– experiment; – model; – model of thermocouple

Figure 9.
Starting diagrams.

fact that the real combustion chamber ignition was some earlier than it is declared in the technical requirements to ACS, and the model implements these requirements.

9. Conclusion

The section discussed a new method for the synthesis of the mathematical model of engine operation in the sub-idle region. As shown, this model belongs to the class of dynamic models. To ensure the unity of the structure with the model of operating modes, a simplified static model is introduced into the dynamic model, which takes into account the synthesized characteristics of the engine components in the format of direct dependencies between the parameter-argument and the parameters used for the synthesis of ACS.

A mathematical model of the thermocouple was also presented and it was proved to become an integral part of the starting model, due to the large inertia of temperature sensors in the sub-idle.

The starting model can be used in the development of control laws for the starting, dead cranking, and in-flight starting.

Acknowledgements

This work has been carried out with the support of the Ministry of Education and Science of Ukraine (research project no. D203-3/2019-П).

Nomenclature

c_p	specific heat capacity at constant pressure condition
J	moment of inertia
L	specific work
M	torque
N	rotation speed
P	pressure
Pow	power
T	temperature
W	air (gas) flow rate
Δ	absolute deviation
ε	thermocouple time constant
η	efficiency
π	pressure ratio
τ	engine time constant

Indexes

a	air
C	compressor
cor	corrected parameter
f	fuel
HP	high-pressure cascade
in	inlet
IP	intermediate-pressure cascade
LP	low-pressure cascade
res	resistance
st	static
T	turbine
TC	thermocouple

Author details

Sergiy Yepifanov* and Feliks Sirenko
National Aerospace University "Kharkiv Aviation Institute", Kharkiv, Ukraine

*Address all correspondence to: aedlab@gmail.com

References

[1] Gaudet SR, Gauthier JED. A simple sub-idle component map extrapolation method. In: ASME Paper No. GT2007–27193. 2007. 9 p

[2] Sexton WR. A method to control turbofan engine starting by varying compressor surge valve bleed [thesis]. Virginia: Virginia Polytechnic Institute and State University; 2001

[3] Bahlawan H, Morini M, Pinelli M, Ruggero Spina P, Venturini M. Development of reliable NARX models of gas turbine cold, warm and hot start-up. In: ASME paper GT2017–63332. 2017. 14 p

[4] Kurzke J. Correlations hidden in compressor maps. In: ASME Paper No. GT2011–45519. 2011. 11 p

[5] Walsh PP, Fletcher P. Gas Turbine Performance. Oxford, UK: Blackwell Science; 1998 646 p

[6] Benzer WA, Johnsen IA, Bullock RO. Compressor operation with one or more blade rows stalled. In: NASA Aerodynamic Design of Axial-Flow Compressors. Vol. SP-36. 1965. pp. 341-346

[7] Kim S, Ellis S, Challener M. Real-time engine modeling of a three shafts turbofan engine: from sub-idle to max power rate. In: ASME Paper No. GT2006–90656. 2006. 7 p

[8] Agrawal RK, Yunis M. A generalized mathematical model to estimate gas turbine starting characteristics. Journal of Propulsion and Power. 1982;**104**: 194-201

[9] Kong C, Ki J, Kang M. A new scaling method for component maps of gas turbine using system identification. Journal of Engineering for Gas Turbines and Power. 2003;**125**(4):979-985

[10] Bretschneider S, Reed J. Modeling of start-up from engine-off conditions using high fidelity turbofan engine simulations. In: ASME Paper GT2015–43528. 2015. 8 p

[11] Ferrand A, Bellenoue M, Bertin Y, Cirligeanu R, Marconi P, Mercier-Calvairac F. High fidelity modeling of the acceleration of a turboshaft engine during a restart. In: ASME Paper GT2018–76654. 2018. 11 p

[12] Jones G, Pilidis P, Curnock B. Extrapolation of compressor characteristics to the low-speed region for sub-idle performance modelling. In: ASME Turbo Expo 2002: Power for Land, Sea, and Air, American Society of Mechanical Engineers. 2002. pp. 861-867

[13] Pourfarzaneh H, Hajilouy-Benisi A, Farshchi M. An analytical model of a gas turbine components performance and its experimental validation. In: ASME Paper No. GT2010–23369. 2010. 6 p

[14] Sukhovii SI, Yepifanov SV, Pavluk EV, Kulik TV. Design-experiment method of GTE starting process linear dynamic simulator forming. In: Aerospace Technique and Technology. Vol. 9/56. Kharkiv: National Aerospace University; 2008. pp. 147-151

Chapter 3

Gas Turbine Simulation Taking into Account Dynamics of Gas Capacities

Sergiy Yepifanov and Roman Zelenskyi

Abstract

The chapter considers one of the main dynamic factors of the turbine engine—the dynamics of gas capacities. Typically, the most influencing capacities in the turbine engine are combustion chamber, afterburner, mixing chamber, secondary duct of turbofan, and jet nozzle. Simulation of high-frequency transients in turbine engines needs taking into account this factor. For the needs of automatic control and parametric diagnostics, the equations of capacities must be combined with the equations of rotor dynamics and, sometimes, with the equations of a measurement system and actuators. The model complexity consists in two features. The first feature is in how many segments are used to simulate the capacity. The second feature is in what of three basic laws are taken into account at the gas motion description: the mass conservation law, the energy conservation law, and the momentum conservation law. This chapter includes the analysis of models of different complexity followed by the conclusions about their applicability. In the last part of the chapter, the real case of the engine dynamics analysis is considered when the designer does not need the simulation of the capacities' dynamics in time, but needs estimating of the capacities' ability to oscillate and in their natural oscillation frequencies.

Keywords: turbine engine dynamics simulation, gas capacities, differential equations, linearization, Eigen frequency

1. Introduction

Engine development is known to include numerous stages and, among them, control systems and engine health management systems development. The development of these systems, however, includes conducting much experimental work, which is not a good choice, keeping in mind the cost of tests and time expenses. An alternative choice to be made is the involvement of the mathematical modeling into a development process [1]. One of the topics to be discussed in this chapter deals with the problem of considerable pneumatic volumes of a gas turbine: main and afterburning combustion chambers, bypass, exhaust nozzle, transition ducts, etc. Stationary gas turbines have extra volumes that must be considered when building up the mathematical model: intake with an air purification system and stack with a noise suppression device. In the models of a gas turbine-driven natural gas pumping

IntechOpen

compressor, the volumes of interest are input and output manifolds as well as a main pipeline. In all abovementioned cases, the designers must allow for the dynamics of pneumatic volumes when studying the transient behavior of an isolated engine or the engine as a component of a power plant.

The design process engages numerous kinds of models depending on the stage the development is at and the problems to be resolved by the particular model. But the "mother" of each and every model is a nonlinear component level model (thermodynamic model), which describes gas path variables using thermodynamic relations and performance maps of all main engine components, such as compressors, turbines, combustors, and input and output devices. When looking carefully at the transients described by this kind of models, one can conclude that the main factors affecting the engine transients are the inertia of rotors, the thermal inertia of engine parts, and the inertia of pneumatic volumes. Usually, designers are good to go with the model that eliminates the last two factors and considers only the rotor dynamics. Nevertheless, for some cases, it is good to have a model that is able to carefully simulate the processes, taking place in the pneumatic volumes, and the phenomenon of thermal inertia. Many researchers have paid their attention to the problem of pneumatic volumes within a total thermodynamic model. Next you will find a brief overview of their findings.

Almost five decades ago, Fawke and Saravanamuttoo proposed a method to simulate gas turbine inter-component volumes within the thermodynamic model [2, 3]. The proposed method found its niche in the field, but it has an increasing degree of differential equations and is too complicated for real-life calculations. The equations that describe the volumes in this method are also known for their low robustness. To cope with these challenges, the designer that is up to use this must take many assumptions to cope with these challenges. In this case, the equations can be simplified, which makes the solution more robust. But, unfortunately, the above studies do not provide any recommendations about a proper algorithm on how to compile the list of requirements for a particular engine. It is not even known whether the dynamics of all inter-component volumes should be simulated.

A few years later, a pretty complete analysis of the general problem of pneumatic volume simulation was made by Glikman in [4]. He described many methods to simulate the volume effect, but, unfortunately, did not pay enough attention to their comparison and highlighting the use cases. Moreover, the book does not cover the specific features of gas turbine engine simulation.

For the past two decades, many scientists have turned their sight to the problem of pneumatic volumes and its effect on engine transients [1, 5–11]. Most of the papers consider the pneumatic volume consisting of a single region with the performance described by the set of differential equations of mass and energy conservation. Conservation laws are added to the equation sets that describe the operation of components. The resulting set of the conservation laws and the equations of components' operation compile the final set of equations, known as thermodynamic model. However, some of the works mentioned above still consider an isothermal process in the volume. Thus, only the pressure alternation is simulated keeping the temperature constant.

Wrapping up the analysis made above, the thermodynamic models including the models of pneumatic volumes take the assumptions listed below:

- In many cases the process in the volume is considered to be isothermal.
- The momentum conservation is omitted.
- The volume model can be called an "all-in-one-volume."

One of the shortcomings of the methods overviewed before consists in omitting the momentum conservation law [12]; however, it plays a considerable role in an overall accuracy of the simulation.

The importance of the momentum conservation was also confirmed by Shi et al. in the study [13]. The authors deal with three engine models called "no volume effect," "traditional simplified volume effect," and "compressibility volume effect." However, this study focuses only on the time delay of the transient because of the volume effect. The other parameters of the transient were not discussed in the study.

The gas turbine models allowing for the momentum conservation in pneumatic volumes are also presented in papers [14–16]. However, the authors do not indicate the range of tasks when it is essential to take into consideration the volume effect.

A new software PROOSIS for simulation in the area of propulsion allows more precise pneumatic volume description that includes the momentum conservation law. Henke et al. in their paper [17] introduce the PROOSIS capabilities in simulating gas turbine transients. One of their conclusions is that the time of transients caused by the volume effect is generally determined by the mass conservation. This is disputable, and it will be shown in the present chapter that indeed the transients are longer.

Wrapping up the overview of the existing methods to simulate the dynamics of the gas path with pneumatic volumes, one can draw the following conclusions:

- These volumes cause the delay of the transients.
- Most of the methods consider the phenomena of mass, energy, and momentum conservation, as well as hydraulic volume resistance; however, no method covers these phenomena simultaneously. Moreover, the used combinations of some phenomena are not compared.
- When the pneumatic volume algorithm is employed for control system design, it must be able to satisfy some specific needs related to the volume, such as oscillation analysis, time response evaluation, and analysis of the natural frequencies. The above tasks should be performed with minimum computation time expenses and, if possible, without simulation of the whole engine. However, none of the overviewed methods can solve these tasks.
- The authors of the overviewed studies did not focus on making faster the engine dynamic model with the algorithms describing pneumatic volumes.

The present chapter aims to overcome the mentioned bottlenecks and propose the best model of pneumatic volumes to be used as a component of either an engine dynamic model or an autonomous usage. Section 1 determines the set of differential equations to generally describe a pneumatic volume. This section introduces seven alternative volume models that adopt different simplifications. Section 2 provides numerical simulation of the volume by each model and, using the simulation results, the models' comparison. The linearization algorithms are described in Section 3, and the linearization accuracy is studied in Section 4.

2. Mathematical models of the pneumatic volumes

2.1 Basic equations

To make a fast computational algorithm, the present approach assumes that a pneumatic volume has a constant transversal section. The proposed thermodynamic model deals with the gas path variables averaged over the radius and the circumference, i.e., it considers the averaged variables of a one-dimensional gas flow. Because most of the methods cited in the introduction use the total pressure and total temperature to characterize the flow (see [1–3, 6–12, 14–17]), the same variables are employed in the present study. In the present approach, the difference between total and static parameters is ignored as the subject of the study is low-speed flow ($M < 0.4$). For the same reason, static density is determined by the total pressure and temperature in the ideal gas state equation $p = \rho RT$.

Let us introduce the volume of interest by **Figure 1** and formulate the conservation laws for this volume. The mass conservation can be presented by

$$\frac{dm}{dt} = W_{in} - W_{out} \tag{1}$$

Let us then use the ideal gas equation and express the mass as $m = \rho V = \frac{P}{RT} LA$, whence the mass conservation law transforms to

$$\frac{dm}{dt} = \frac{LA}{R}\left(\frac{1}{T}\frac{dP}{dt} - \frac{P}{T^2}\frac{dT}{dt}\right) \tag{2}$$

The internal energy conservation law for an adiabatic flow is written as

$$\frac{dU}{dt} = h_{in}W_{in} - h_{out}W_{out}. \tag{3}$$

After expressing the internal energy by the pressure, one obtains the energy conservation law that reflects the differential equation for the pressure:

$$U = mc_vT = \frac{LAc_v}{R}P, \tag{4}$$

$$\frac{dU}{dt} = \frac{LAc_v}{R}\frac{dP}{dt}, \tag{5}$$

$$\frac{dP}{dt} = \frac{\gamma R}{LA}(T_{in}W_{in} - TW_{out}) \tag{6}$$

Let us now get the differential equation for the temperature using Eqs. (6) and (2):

$$\frac{dT}{dt} = \frac{RT^2}{PLA}\left[\left(\gamma\frac{T_{in}}{T} - 1\right)W_{in} - (\gamma - 1)W_{out}\right] \tag{7}$$

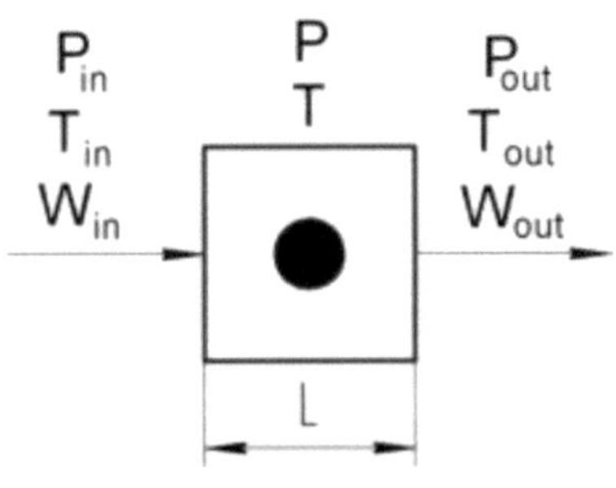

Figure 1.
Gas capacity design scheme.

The next equation to form presents the momentum conservation law expressed through the Darcy-Weisbach equation (it relates the pressure loss due to friction to the average velocity of the flow for an incompressible fluid):

$$\Delta P = \xi \frac{RT}{2PA^2} W^2. \tag{8}$$

It helps to describe the relations between the variables of two elementary volumes situated to the left and to the right from the calculation point:

$$\frac{dW_{in}}{dt} = \frac{2A}{L}\left(P_{in} - P - \xi \frac{RT_{in}}{2P_{in}A^2} W_{in}^2\right); \tag{9}$$

$$\frac{dW_{out}}{dt} = \frac{2A}{L}\left(P - P_{out} - \xi \frac{RT}{2PA^2} W_{out}^2\right), \tag{10}$$

where $\xi = \varsigma \frac{L}{2D}$ is the Darcy friction factor, ς is the specific frictional resistance, $D = \frac{4A}{\Pi}$ is the hydraulic diameter, and Π is the wetted perimeter of the cross section.

Eqs. (6), (7), (9) and (10) constitute a closed set of first-order nonlinear differential equations. The set describes four parameters P, T, W_{in}, and W_{out} that characterize the volume dynamics.

As proven before, the processes in the gas path pneumatic volumes can be simulated with different precision.

The simplest method only regards the mass conservation (see [18]). According to this method, the time derivative of temperature in Eq. (2) is considered negligibly small. Modification of this approach is used in the paper [19] for compressor dynamics simulation. The method described by Shevyakov [20] suggests the energy conservation to be considered only when deducing the set of equations. The problem of calculation accuracy is omitted. The method used by Dobryansky [21] already considers the mass and energy conservation. One of last publications on volume dynamic modeling [22] is based on the same suppositions. However, the relation between the internal energy and the temperature employs the heat capacity at constant pressure instead of the constant volume heat capacity. The method considered by Jaw and Mattingly [1] is also based on mass and energy conservation, but Eq. (2) has no time derivative of temperature. The method described by Gurevich [23] already takes into consideration the difference between the static and total parameters but still neglects the momentum conservation. Such diversity of the methods for simulating the pneumatic volumes of gas turbines implies the necessity to perform their comparative study.

The present chapter introduces and compares three groups of pneumatic volume models.

The first group unites all isothermal models. The assumption about the minuteness of the second item on the right side of Eq. (2) is equivalent to an assumption about the isothermal process in the volume. Keeping the temperature constant requires heat exchange with the ambiance, and hence the volume process cannot be adiabatic in this case. Indeed, the volumes in real engines are not absolutely adiabatic because the heat exchange is always present between the working substance and the construction elements surrounding the cavity. However, the characteristic time of the heat exchange is several orders greater than that of the mass and energy accumulation in the volumes. This fact proves the use of the adiabatic models. Although the isothermal models are not the best option for the volume effect simulation, the present chapter uses them for comparing the errors of different models.

The second group includes the models based on the mass and energy conservation. The volume process in these models is adiabatic.

The third group consists of adiabatic models considering all conservation laws (mass, energy, and momentum).

2.2 Model 1.1: Isothermal volume without hydraulic resistance

Given the assumptions that are taken for this model $\xi = 0$, $\frac{dT}{dt} = 0$, it follows that $T = T_{in}$-const.

Let us differentiate Eq. (6) and then substitute the flow rate time derivatives that correspond to the case $\xi = 0$. As a result, we have

$$\tau_1^2 \frac{d^2P}{dt^2} + P = \frac{1}{2}(P_{in} - P_{out}), \tag{11}$$

where $\tau_1 = \frac{1}{2}\tau_0$, $\tau_0 = \frac{L}{a}$ is a time required for the disturbance to pass through the volume, and $a = \sqrt{\gamma RT}$ is the sonic velocity in the cavity.

Thus, the lossless isothermal volume is modeled by a single second-order linear differential equation, whose solution depends on input disturbances and the time constant τ_1.

2.3 Model 1.2: Isothermal volume with hydraulic resistance (momentum conservation is omitted)

Having applied the condition $\frac{dW_{in}}{dt} = \frac{dW_{out}}{dt} = 0$ for Eqs. (9) and (10), we get the following:

$$W_{in} = A\sqrt{\frac{2P_{in}(P_{in} - P)}{\xi RT_{in}}};\ W_{out} = A\sqrt{\frac{2P(P - P_{out})}{\xi RT}}. \tag{12}$$

Then, the Eq. (6) is transformed to

$$\frac{dP}{dt} = \frac{\gamma}{L}\sqrt{\frac{2RT}{\xi}}\left[\sqrt{P_{in}(P_{in} - P)} - \sqrt{P(P - P_{out})}\right]. \tag{13}$$

Thus, in the case of isothermal volume with hydraulic losses, the volume is described by a first-order nonlinear differential equation.

2.4 Model 1.3: Isothermal volume with hydraulic resistance (momentum conservation is taken into account)

The volume is modeled by the following set of equations consisting of Eq. (6) modified for the constant temperature condition and Eqs. (9) and (10):

$$\frac{dP}{dt} = \frac{\gamma RT}{LA}(W_{in} - W_{out}); \tag{14}$$

$$\frac{dW_{in}}{dt} = \frac{2A}{L}\left(P_{in} - P - \xi\frac{RT_{in}}{2P_{in}A^2}W_{in}^2\right); \tag{15}$$

$$\frac{dW_{out}}{dt} = \frac{2A}{L}\left(P - P_{out} - \xi\frac{RT}{2PA^2}W_{out}^2\right). \tag{16}$$

2.5 Model 2.1 based on mass and energy accumulation in the volume without hydraulic resistance

As momentum loss is neglected, we can state that $\frac{dW_{in}}{dt} = \frac{dW_{out}}{dt} = 0$. Since the pressure loss in Eqs. (9) and (10) is equal to zero, we have $P = P_{in} = P_{out}$. But then it follows from Eq. (1) that $\frac{dm}{dt} = 0$, and we can transform Eq. (7) to

$$\frac{dT}{dt} = \frac{\gamma RWT^2}{PLA}\left(\frac{T_{in}}{T} - 1\right). \tag{17}$$

2.6 Model 2.2 based on mass and energy accumulation in the volume with hydraulic resistance

By substituting Eq. (12) into Eq. (6) and Eq. (7), we arrive to the following equations for the model under analysis:

$$\frac{dP}{dt} = \frac{\gamma T}{L}\sqrt{\frac{2R}{\xi}}\left[\sqrt{\frac{P_{in}(P_{in} - P)}{T_{in}}} - \sqrt{\frac{P(P - P_{out})}{T}}\right]; \tag{18}$$

$$\frac{dT}{dt} = \frac{T^2}{P}\frac{1}{L}\sqrt{\frac{2R}{\xi}}\left[\begin{array}{l}\left(\gamma\frac{T_{in}}{T} - 1\right)\sqrt{\frac{P_{in}(P_{inl} - P)}{T_{in}}} - \\ -(\gamma - 1)\sqrt{\frac{P(P - P_{out})}{T}}\end{array}\right]. \tag{19}$$

2.7 Model 3.1 based on mass and energy accumulation and momentum conservation in the volume without hydraulic resistance

The set of equations, constituting this model, consists of Eqs. (6) and (7), and also Eqs. (9) and (10) changed for the lossless conditions ($\xi = 0$):

$$\frac{dW_{in}}{dt} = \frac{2A}{L}(P_{in} - P); \tag{20}$$

$$\frac{dW_{out}}{dt} = \frac{2A}{L}(P - P_{out}). \tag{21}$$

This model contains a contradiction that can be illustrated by the following example. The change of pressure in a volume inlet results in the pressure drop between the volume inlet and outlet. So the pressure drop in its turn makes the gas flow to become transient (see Eqs. (20) and (21)). Since the volume of interest is lossless, the pressure at the inlet and outlet will eventually become equal when the transient comes to the steady state. However, this will never happen as the inlet pressure has already changed, and the outlet pressure will remain immutable forever. Thus, the transient will not stop within this model.

Despite the above contradiction, the considered model is not expelled from the study, because it still can be used autonomously to estimate the dynamic process in the volume.

2.8 Model 3.2 based on mass and energy accumulation and momentum conservation in the volume with hydraulic resistance

As it has been mentioned above, this model is the most comprehensive. It is based on Eqs. (6), (7), (9) and (10).

3. Simulation

The language VisSim was used for programming the algorithm of volume dynamic simulation with the above models. To ensure the accuracy, the simulation was performed by different integration algorithms (Euler, Runge Kutta 2nd order, Runge Kutta 4th order, adaptive Runge Kutta 5th order) and with different integration steps. The simulation results were trusted only when all integration algorithms provided similar results and the integration step was small enough not to influence them.

As proposed in the paper [12], to simulate the transients, we have chosen the volume with standard geometrical characteristics L = 1 m and A = 1 m^2. This volume is placed between two infinite capacities, whose parameters are P_{in} = 300 kPa, P_{out} = 150 kPa, and T_{in} = T_{out} = 300 K. The first capacity was simulated as a single volume. Having experimented with different integration technics and diverse integration steps, we arrived to proper computation conditions at which the integration method and integration step do not influence simulation results. Under these conditions, the computations were conducted with the seven models described above. The results are plotted in **Figure 2** (the disturbing factor is a pressure drop at the inlet ΔP = 10 kPa) and **Figure 3** (the disturbing factor is a temperature drop at the inlet ΔT = 50 K).

Using these figures, let us firstly analyze the dynamic performance of each model and then study the effect of volume split-off on simulated parameters.

3.1 Model dynamic performances

Since Model 3.2 is the most comprehensive, we will employ its performance as a pattern to compare the performances of the other models with it.

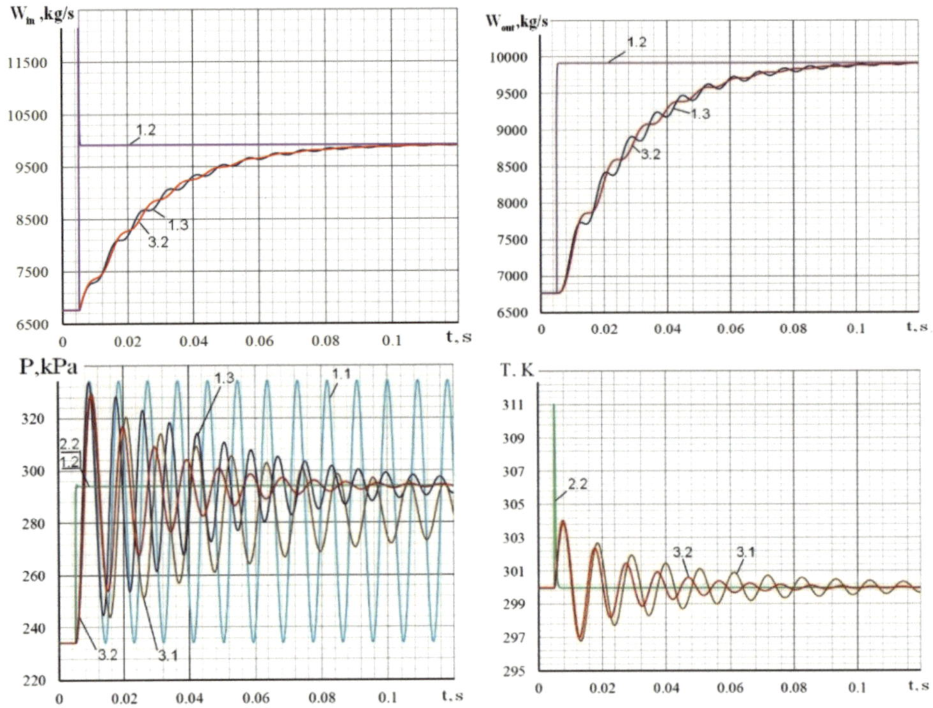

Figure 2.
Reaction of the volume parameters on the perturbation in the inlet pressure (single volume model).

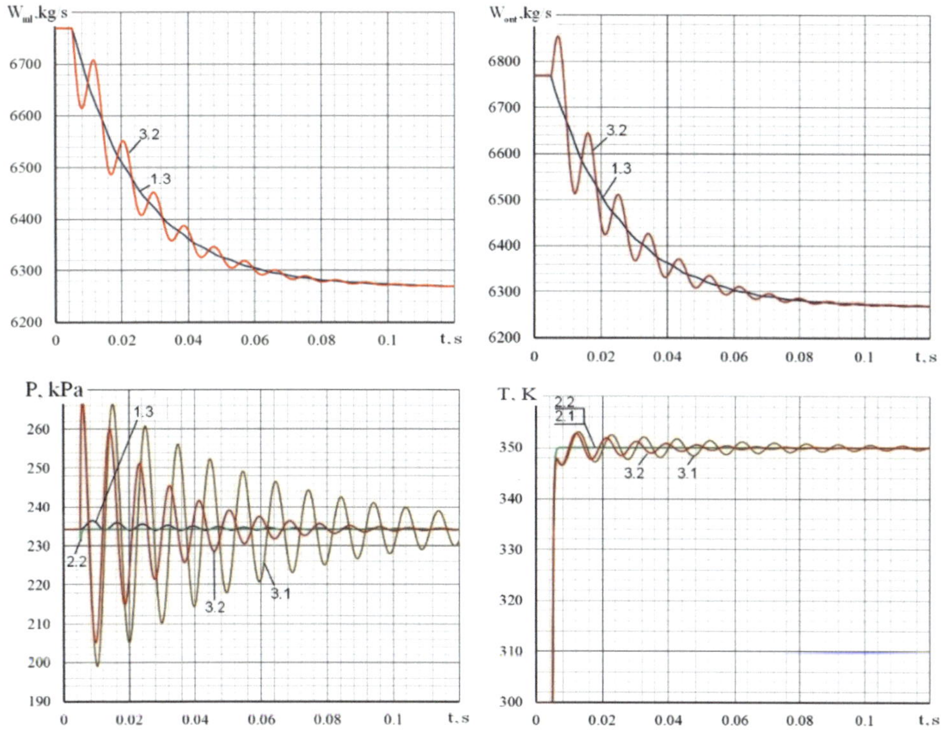

Figure 3.
Reaction of the volume parameters on the perturbation in the inlet temperature (single volume model).

Plots of model 1.1 express the results in the constant-amplitude continuous oscillations. The model response has this form because Eq. (11) is similar to a conservative element [24]. Thus, this model cannot be used to properly simulate the volume effect within the gas path model. However, this model represents the frequency of the parameter oscillations well enough and can be used in a volume Eigen frequency analysis.

Models 1.2 and 2.2 give the similar results when simulating the pressure. The transient lags in this case. Its time constant is very small (about 0.002 s), much smaller than the total transient simulation time. Models 2.1 and 2.2 output a very similar response when simulating the temperature (however, as mentioned above, model 2.1 cannot simulate the pressure).

The pressure simulation using models 1.2., 1.3, 2.2, and 3.2 for the case of an inlet pressure perturbation gives the same values to the end of the transient. The pressure transient computed by model 3.1 ends with the different value, which is equal to the average between the inlet and outlet pressure. The difference appears because models 1.2., 1.3, 2.2, and 3.2 consider hydraulic losses in contrast to model 3.1.

The parameters simulated by models 1.3, 3.1, and 3.2 change according to a damped oscillation law. Obviously, this is because these models take into account the momentum conservation law.

As shown in the figures, model 1.3 has a bit higher frequency of oscillations than model 3.2. In general, the frequencies of different models are close to each other. Hence, all these models can be used when estimating the amplitude-frequency characteristic of the volume.

As regards the oscillation decay time, it is two times greater for model 1.3 and five times greater for model 3.1 than the model 3.2 time.

Let us now consider the effect of volume split-off on the simulated parameters.

3.2 Volume split-off effect simulation

The most advanced model 3.2 has been chosen to study this effect. Three cases were considered: the entire cavity not split, the cavity split in the axial direction into three equal volumes, and the cavity split into five volumes. The section area remained the same.

The presented above equations, corresponding to the model chosen, were applied to each one elementary volume. The output parameters (pressure, temperature, and flow rate) of one volume were the input parameters of the next volume. The specific frictional resistance ξ of each elementary volume was determined in the way that results in the total pressure loss equal to that of the non-split cavity.

The simulation results for the three cases are plotted in **Figures 4** and **5**. As seen in these figures, the transient plots corresponding to these cases are pretty similar, i.e., all of them obey the damped oscillation law. When the pressure disturbance is considered (**Figure 4**), the rate of the damping is greater for the three-volume and five-volume models than for the single volume model. For the temperature disturbance, the damping rates of all the models are equal. The fundamental frequencies for different volume numbers are very close as well. As to the amplitude, the three-volume and five-volume models have approximately equal amplitudes that are about 20% greater than that of the single volume model.

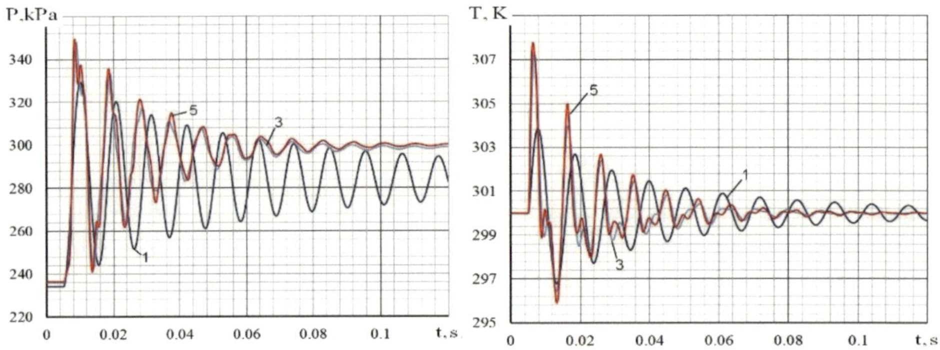

Figure 4.
Reaction of the volume parameters on the perturbation in the inlet pressure: (1, ▬)—Single volume, (3, ▬)—Three volumes, and (5, ▬)—Five volumes.

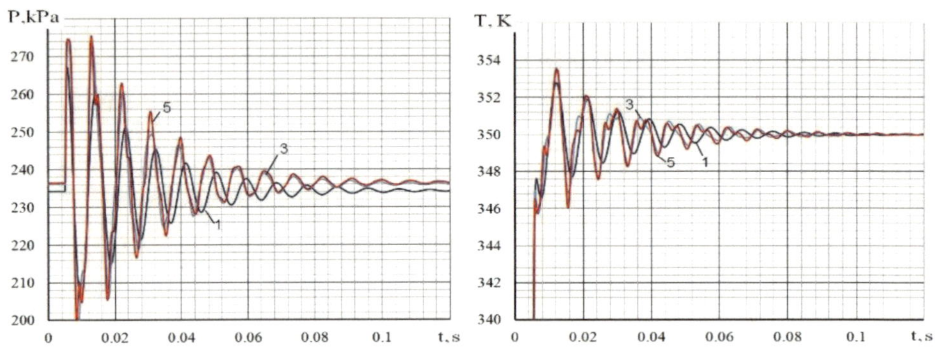

Figure 5.
Reaction of the volume parameters on the perturbation in the inlet temperature: (1, ▬)—Single volume, (3, ▬)—Three volumes, and (5, ▬)—Five volumes.

4. Linear analysis of volume effect

The differential equations of the models described in the previous section can be generally presented by

$$\frac{dy}{dt} = f(x_1, \dots, x_n, y). \tag{22}$$

The nonlinear form of the equations does not allow the direct use of the universal methods [24, 25] that have been specially developed for the dynamic analysis of the set of linear differential equations. Hence, let us linearize the equations describing the volume effect and transform them introducing small variations of arguments from their steady-state magnitudes denoted by the subscript "0."
We arrive to

$$\frac{d(\Delta y)}{dt} = \frac{df}{dx_1}\Delta x_1 + \dots + \frac{df}{dx_n}\Delta x_n + \frac{df}{dy}\Delta y, \tag{23}$$

where $\Delta x_i = x_i - x_{i\,0}$ and $\Delta y = y - y_0$. The time derivatives were evaluated here in the point $(x_{1\,0}, \dots, x_{n\,0}, y_0)$.

Let us transform absolute deviations to relative deviations $\delta y = \frac{\Delta y}{y_0}$, $\delta x_1 = \frac{\Delta x_1}{x_{1\,0}}, \dots,$ $\delta x_n = \frac{\Delta x_n}{x_{n\,0}}$. Eq. (23) is then changed to

$$y_0 \frac{d(\delta y)}{dt} = \frac{\partial f}{\partial x_1} x_{1\,0}\delta x_1 + \dots + \frac{\partial f}{\partial x_n} x_{n\,0}\delta x_n + \frac{\partial f}{\partial y} y_0 \delta y. \tag{24}$$

Using the linearization principle described above, linear differential equations have been formed for all the models under consideration, and their analytical solutions for volume pressure and temperature were derived (see Appendix). These solutions are determined by the totality of the physical laws and assumptions that are used in each of the considered models.

Let us now analyze basic properties of these solutions. Specifically, in the next section we will determine the order of equations, their parameters, and type of transients that they describe.

5. Analysis of the transients in the volume based on the linearized equations of each model

5.1 Model 1.1

Solution (51) of this model corresponds to the undamped harmonic oscillations with the angular frequency $\omega = \frac{1}{\tau_1}$. When L = 1 m and a_0 = 500 m·s^{-1}, then τ_1=0.001 s.

5.2 Model 1.2

Eq. (55) corresponds to an aperiodic system, whose dynamics is described by the time constant τ_p.

For the rough estimation of the time constant and the gain coefficient, we can neglect the Darcy friction factor and assume that $\frac{P_{in\ 0}}{P_0} \approx \frac{P_{out\ 0}}{P_0} \approx 1$. As a result, we arrive to

$$\tau_P = \frac{1}{2}\xi\tau_0 M_0; K_{in}^P = \frac{1}{2}; K_{out}^P = \frac{1}{2}. \tag{25}$$

Provided that L = 1, ξ = 0.02, and a = 500 ms^{-1}, the time constant is τ_P = 0.00006 s.

It is obvious that the model simulates the volume like an almost inertia-free object.

5.3 Model 1.3

Let us analyze Eq. (61) that presents this model. The aperiodicity conditions can be given by an inequality:

$$\left(\frac{\tau_P}{1 + 0.25\xi\gamma M_0^2}\right)^2 > 4\frac{\tau_0^2}{2 + 0.5\xi\gamma M_0^2}. \tag{26}$$

The Darcy friction factor and squared Mach number are minuscule. Hence, this condition can be simplified to

$$\tau_P^2 > 2\tau_0^2 \text{ or } \xi^2\tau_0 M_0^2 > 8. \tag{27}$$

As we see, this condition is not fulfilled. Thus, the dynamic processes in the volume have an oscillatory nature. The coefficients in the right side of Eq. (61) are positive. Hence, the system is robust, i.e., the oscillations relax. The time constant τ_2 determines the intensity of the relaxation. It is inverse to the real root α of the characteristic equation corresponding to differential Eq. (61):

$$\tau_2 = \frac{1}{\alpha} = -\frac{2\tau_0^2}{\tau_P}\frac{1 + 0.25\xi\gamma M_0^2}{2 + 0.5\xi\gamma M_0^2} \approx -\frac{\tau_0^2}{\tau_P} = -\frac{2\tau_0}{\xi M_0} = -\frac{2\tau}{\xi}, \tag{28}$$

where $\tau = \frac{L}{c_0}$ is the time needed by the flow to cross the volume.

The frequency is equal to an absolute value of the imaginary root:

$$\omega = \frac{\sqrt{\frac{\tau_P^2}{\left(1+0.25\xi\gamma M_0^2\right)^2} - \frac{4\tau_0^2}{2+0.5\xi\gamma M_0^2}}}{\frac{2\tau_0^2}{2+0.5\xi\gamma M_0^2}} \approx \frac{\sqrt{2}}{\tau_0} = \frac{\sqrt{2}a_0}{L}. \tag{29}$$

The evaluated time constant is much greater than that from model 1.2. The frequency is $\sqrt{2}$ times lower than the frequency estimated by model 1.1. However, this difference is acceptable for rough estimation.

5.4 Model 2.1

Eq. (64) corresponds to an aperiodic system, which characteristic time of the transient is given by the time constant τ_T.

If $L = 1$ m, $c_0 = 100\ \mathrm{m{\cdot}s^{-1}}$, and the ratio of specific heats is 1.4, then $\tau_T = 0.007$ s. This constant is small, but it is considerably greater than the time constant obtained for model 1.2.

5.5 Model 2.2

The left sides of Eqs. (70) and (71) that represent this model have the same order and similar coefficients. This proves the dynamics of pressure and temperature in the volume to be equal.

The aperiodicity condition for these equations can be formulated as

$$\left(\frac{\tau}{\gamma}+\frac{\gamma+1}{2\gamma}\tau_P\right)^2 > 4\frac{\tau\tau_P}{\gamma}. \tag{30}$$

As $\tau_P = \frac{1}{2}\xi\tau_0 M_0$, the condition $\tau_P << \tau$ is fulfilled, and the aperiodicity condition is transformed to $\tau < 4\gamma\tau_P$. It is obvious that this condition is fulfilled. Hence, the transients have an aperiodic form. The dynamics of the volume is determined by its time constants:

$$\tau_1 \approx \frac{\tau}{\gamma} \text{ and } \tau_2 \approx \tau_P. \tag{31}$$

5.6 Model 3.1

Let us analyze Eqs. (79) and (80) derived for model 3.1. For doing so, we must form a characteristic equation, which is common for both equations:

$$s^3 + as^2 + bs + c = 0, \tag{32}$$

where $a = \frac{\gamma}{\tau}$, $b = \frac{4}{\tau_0^2}$, and $c = \frac{4}{\tau\tau_0{}^2}$.

Let us use the method proposed by Gerolamo Cardano [26]. For this we first check whether the volume dynamics is oscillatory. The condition of the oscillations is $Q>0$, where

$$Q = \left(\frac{p}{3}\right)^3 + \left(\frac{q}{2}\right)^2; p = -\frac{a^2}{3} + b = -\frac{\gamma^2}{3\tau^2} + \frac{4}{\tau\tau_0 M_0}; q = 2\left(\frac{a}{3}\right)^3 - \frac{ab}{3} + c$$
$$= 2\left(\frac{\gamma}{3\tau}\right)^3 - \frac{4\gamma}{3\tau\tau_0^2} + \frac{4}{\tau\tau_0^2}.$$

Since (1) $\tau_0 < \tau$, (2) $\frac{\gamma^2}{3\tau^2} << \frac{4}{\tau_0^2}$ and $\frac{4\gamma}{3\tau\tau_0^2} < \frac{4}{\tau\tau_0^2}$, (3) $p>0$, and (4) $q>0$, the condition $Q>0$ is fulfilled. This obviously means that the transient has an oscillatory character.

The characteristic Eq. (32) has a single real root s_1 and two complex conjugate roots s_2 and s_3:

$$s_1 = A + B - \frac{a}{3} = A + B - \frac{\gamma}{3\tau}; \tag{33}$$

$$s_{2,3} = \alpha \pm i\omega = -\frac{A+B}{2} - \frac{\gamma}{3\tau} \pm i\frac{A-B}{2}\sqrt{3}, \tag{34}$$

where $A = \sqrt[3]{-\frac{q}{2}+\sqrt{Q}}$; $B = \sqrt[3]{-\frac{q}{2}-\sqrt{Q}}$.

Analyzing Eqs. (33) and (34), we can see that they have some infinitesimal summands that can be neglected. In this way, we get the simplified equations:

$$p \approx b = \frac{4}{\tau_0^2}; q \approx -\frac{ab}{3}b + c = \frac{4M_0}{\tau_0^3}\left(1 - \frac{\gamma}{3}\right); Q \approx \frac{1}{\tau_0^6}\left[c + x^2\right] \approx \left(\frac{4}{3\tau_0^2}\right)^3; \tag{35}$$

$$A \approx -\frac{1}{\tau_0}\sqrt[3]{x - \sqrt{c + x^2}}; B \approx -\frac{1}{\tau_0}\sqrt[3]{x + \sqrt{c + x^2}}, \tag{36}$$

where $x = 2M_0\left(1 - \frac{\gamma}{3}\right)$; $c = \left(\frac{4}{3}\right)^3$;

$$s_1 \approx -\frac{1}{\tau_0}\left(\sqrt[3]{x - \sqrt{c + x^2}} + \sqrt[3]{x + \sqrt{c + x^2}} + \frac{1}{3}\gamma M_0\right); \tag{37}$$

$$\alpha \approx -\frac{1}{\tau_0}\left[\frac{1}{3}\gamma M_0 - \frac{1}{2}\left(\sqrt[3]{x - \sqrt{c + x^2}} + \sqrt[3]{x + \sqrt{c + x^2}}\right)\right]; \tag{38}$$

$$\omega \approx \frac{\sqrt{3}}{2\tau_0}\left(\sqrt[3]{x + \sqrt{c + x^2}} - \sqrt[3]{x - \sqrt{c + x^2}}\right). \tag{39}$$

Taking into account the Mach number being less than 0.3, these formulas for the characteristic equation roots can be further simplified. Let us change the equations to linear relations of the following form:

$$y = y(M_0 = 0) + \frac{dy}{dM}(M = 0)M_0. \tag{40}$$

The required derivatives are.

$$\frac{ds_1}{dM}(M_0 = 0) = -\frac{1}{\tau_0}; \frac{d\alpha}{dM}(M_0 = 0) = -\frac{\gamma - 1}{2\tau_0}; \frac{d\omega}{dM}(M_0 = 0) = 0, \tag{41}$$

that is why,

$$s_1 \approx -\frac{M_0}{\tau_0} = -\frac{1}{\tau}; \alpha \approx -\frac{\gamma - 1}{2\tau}; \omega \approx \frac{1}{\tau_0}. \tag{42}$$

5.7 Model 3.2

The differential Eqs. (88) and (89) of this model have the common characteristic Eq. (32), where

$$a = \frac{\gamma + 4\xi}{\tau}, b = \frac{2}{\tau_0^2}\left[2 + (2\gamma + 1)\xi M_0^2\right], \text{and } c = \frac{2C}{\tau{\tau_0}^2}. \tag{43}$$

The following parameters were evaluated by the Cardano's method:

$$p = -\frac{a^2}{3} + b \approx \frac{2}{\tau_0^2}\left[2 + (2\gamma + 1)\xi M_0^2\right], q = 2\left(\frac{a}{3}\right)^3 - \frac{ab}{3} + c \approx \frac{4M_0}{3\tau_0^3}(3 - \gamma - 4\xi), \tag{44}$$

$$Q = \left(\frac{p}{3}\right)^3 + \left(\frac{q}{2}\right)^2 \approx \left\{\frac{2}{3\tau_0^2}\left[2 + (2\gamma + 1)\xi M_0^2\right]\right\}^3; \tag{45}$$

$$A \approx -\frac{1}{\tau_0}\sqrt[3]{x - \sqrt{c + x^2}}; B \approx -\frac{1}{\tau_0}\sqrt[3]{x + \sqrt{c + x^2}}, \tag{46}$$

where $x = \frac{2}{3}M_0(3 - \gamma - 4\xi)$; $c = \left\{\frac{2}{3\tau_0^2}\left[2 + (2\gamma + 1)\xi M_0^2\right]\right\}^3$.

The expressions for the parameters of the characteristic equation roots are similar to that of Eqs. (37) and (38) (for model 3.1):

$$s_1 \approx -\frac{1}{\tau_0}\left(\sqrt[3]{x - \sqrt{c + x^2}} + \sqrt[3]{x + \sqrt{c + x^2}} + \frac{1}{3}(\gamma + 4\xi)M_0\right); \qquad (47)$$

$$\alpha \approx -\frac{1}{\tau_0}\left[\frac{1}{3}(\gamma + 4\xi)M_0 - \frac{1}{2}\left(\sqrt[3]{x - \sqrt{c + x^2}} + \sqrt[3]{x + \sqrt{c + x^2}}\right)\right]. \qquad (48)$$

Eq. (39) for ω remains the same.

In the same way as it was done for Eq. (78), the linearization of these expressions over the Mach number allows their simplification:

$$s_1 \approx -\frac{1}{\tau}; \alpha \approx -\frac{\gamma - 1 + 4\xi}{2\tau}; \omega \approx \frac{1}{\tau_0}. \qquad (49)$$

It follows from Eq. (49) that the hydraulic resistance of the volume mostly affects the oscillation damping rate but does not influence the frequency and the aperiodic component of the transient.

5.8 Simulation results

To verify that the linearization did not introduce big error and the obtained results can be trusted, we have compared them with the original nonlinear models. **Figures 6** and 7 illustrate this comparison by plotting pressure for the transients

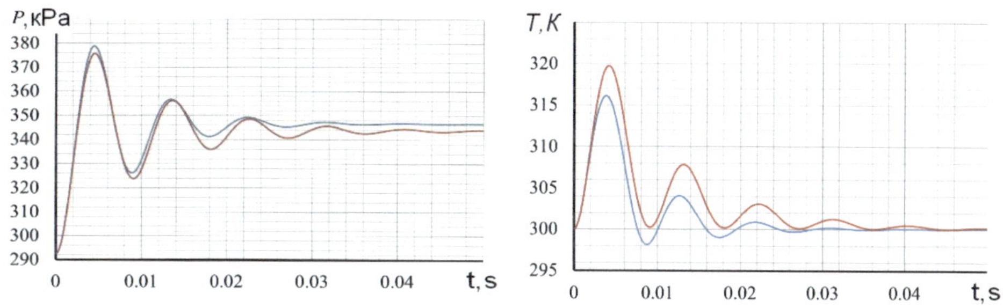

Figure 6.
Reaction of the volume pressure on the perturbation in the inlet pressure (single volume cavity, ▬ nonlinear model, ▬ and linear model).

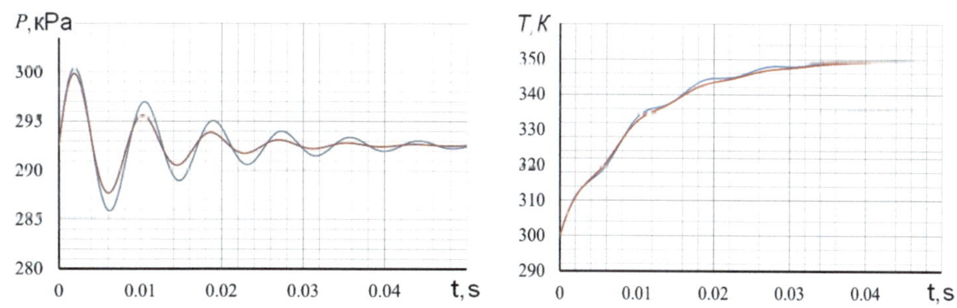

Figure 7.
Reaction of the volume pressure on the perturbation in the inlet temperature (single volume cavity, ▬ nonlinear model, and ▬ linear model).

Model	Transient	Time constants		Eigen frequency ω	
		Formula	Value, s	Formula	Value, 1/s
1.1	Oscillatory	∞	∞	$\frac{2}{\tau_0}$	1000
1.2	Aperiodic	$\tau_P \approx \frac{1}{2}\xi\tau_0 M_0$	0.000084	—	
1.3	Oscillatory	$\frac{2\tau}{\xi}$	0.0476	$\frac{\sqrt{2}}{\tau_0}$	707
2.1	Aperiodic	$\frac{\tau}{\gamma}$	0.00714	—	—
2.2	Aperiodic	$\frac{\tau}{\gamma}, \tau_P$	0.00714, 0.000084	—	—
3.1	Oscillatory	$\tau, \frac{2\tau}{\gamma-1}$	0.01, 0.05	$\frac{1}{\tau_0}$	500
3.2	Oscillatory	$\tau, \frac{2\tau}{\gamma-1+4\xi}$	0.01, 0.00962	$\frac{1}{\tau_0}$	500

Table 1.
Dynamic parameters of the volume.

caused by inlet pressure and inlet temperature perturbations. As seen in the figures, the linear model correctly simulates the main features and parameters of the transient: oscillatory nature, time of the transient, oscillation frequency, and the magnitude of the first overshoot. It is worth to mention that just these performances are the subject of the dynamics analysis for the development of ACSs. The comparison results for all the models confirm that the dynamic behavior of the linear models agree with the behavior of the nonlinear models. This allows recommending the obtained linear models and corresponding analytical solutions for practical usage.

The dynamic parameters obtained as a result of the linearization are presented in **Table 1** for all the models. The numerical values correspond here to the input conditions of the example: L = 1 m, c = 100 m/s, a = 500 m/s, and ξ = 0.42.

6. Conclusions

The following conclusions can be drawn on the results of the carried-out research:

1. The use of the momentum conservation law makes a tangible contribution in the transient state simulation. Thus, it cannot be omitted, when simulating the engine transients by the engine model with the volume model integrated.

2. The volume effect can be accurately simulated by the single volume model. The simulation of big connected volumes (e.g., annular manifolds of gas pumping units or station or trunk pipelines) requires deeper understanding and further researching to prove the model applicability.

3. The isothermal models are not recommended to be integrated into the gas path models because they do not correspond to the operating conditions in the engines.

4. The time of transients evaluated by the conventional volume models 2.1 and 2.2 is significantly lower against the models that among other consider

momentum conservation. Hence, when this time delay effect is the subject of simulation, it is reasonable to use model 3.2.

5. In some instances, it may become important to study the volume effect on the frequency responses of the engine. In this case, we once more recommend model 3.2, which consists of Eqs. (6), (7), (9) and (10). The momentum transformation in the cavity causes oscillations of the parameters. The frequency of oscillations depends on the velocity of sound and the volume length only.

6. The time of the transients depends on the aperiodic component duration and the oscillation decay time. The aperiodic component duration in its turn depends on the time during which the gas crosses the volume. The decay time may be greater than the time of the aperiodic process.

7. The hydraulic resistance mostly affects the oscillation decay time. High hydraulic resistance reduces it. If the hydraulic losses are negligible, then the oscillation decay time is about five times longer than the aperiodic process duration.

8. The analytical method to solve the equations of volume dynamics makes it possible to determine the main dynamic properties of the volume and to get simple equations for determining the dynamic parameters on the basis of known geometrical characteristics and gas properties.

9. The obtained analytical equations and solutions (Eqs. (88)–(93) are the most accurate) can be implemented when developing the combined algorithm of engine dynamic simulation with the volume effect integrated. Integration of differential equations of the rotor dynamics will be performed iteratively using the required integration step, and the volume effect will be computed analytically. The application of this method will allow significant reduction of the operational time.

Acknowledgements

This work has been carried out with the support of the Ministry of Education and Science of Ukraine (Research Project No. D203-3/2019-П).

Nomenclature

a	velocity of sound
c_v	specific heat capacity at constant volume condition
c	gas velocity
D	hydraulic diameter
A	constant section area
W	gas flow rate
h	enthalpy
γ	ratio of specific heats
L	length of the volume

M	Mach number
m	mass
P	pressure
R	gas constant
T	temperature
U	internal energy
V	volume
ρ	static density
ζ	specific frictional resistance
ξ	friction factor
ω	Eigen frequency
τ	time constant

Indexes

in	inlet
out	outlet
0	initial static value

Appendix

A.1 Linearized equations' derivation

A.1.1 Model 1.1

Differential Eq. (11) is already linear, and thus its structure is conserved despite switching to the relative deviations:

$$\tau_1^2 \frac{d^2(\delta P)}{dt^2} + \delta P = \frac{1}{2}(\delta P_{inl} - \delta P_{out}). \quad (50)$$

The system, which behaves like this, is oscillatory. The solution of this equation in the case of inlet or outlet pressure step is changed by $\delta P = A_P$ must be found as

$$\delta P(t) = \frac{1}{2} A_P \left(1 - \cos \frac{t}{\tau_1}\right). \quad (51)$$

A.1.2 Model 1.2

Let us start from linearizing Eq. (13):

$$\frac{d(\Delta P)}{dt} = \frac{\gamma}{L}\sqrt{\frac{2RT}{\xi}}\left[\frac{(2P_{inl\ 0} - P_0)\Delta P_{inl} - P_{inl\ 0}\Delta P}{2\sqrt{P_{inl\ 0}(P_{inl\ 0} - P_0)}} - \frac{(2P_0 - P_{out\ 0})\Delta P - P_0 \Delta P_{out\ 0}}{2\sqrt{P_0(P_0 - P_{out\ 0})}}\right]. \quad (52)$$

One must consider then

$$W_0 = A\sqrt{\frac{2P_{inl\ 0}(P_{inl\ 0} - P_0)}{\xi RT}} = A\sqrt{\frac{2P_0(P_0 - P_{out\ 0})}{\xi RT}}, \quad (53)$$

whence

$$P_0 = \sqrt{P_{inl\ 0}^2 + \left(\frac{P_{inl\ 0} - P_{out\ 0}}{2}\right)^2} - \frac{P_{inl\ 0} - P_{out\ 0}}{2}. \quad (54)$$

Then, when switching to the relative deviations, we get

$$\tau_P \frac{d(\delta P)}{dt} + \delta P = K_{inl}^P \delta P_{inl} + K_{out}^P \delta P_{out}, \quad (55)$$

where $\tau_P = \frac{P_0}{P_{inl\ 0} + 2P_0 - P_{out\ 0}} \xi \tau_0 M_0$ is a time constant; $M_0 = \frac{c_0}{a}$ is a Mach number; and $K_{inl}^P = \frac{2P_{inl\ 0} - P_0}{P_{inl\ 0} + 2P_0 - P_{out\ 0}} \frac{P_{inl\ 0}}{P_0}$ and $K_{out}^P = \frac{P_{out\ 0}}{P_{inl\ 0} + 2P_0 - P_{out\ 0}}$ are gain coefficients.

The solution of this equation in the case of inlet pressure perturbation $\delta P_{inl} = A_P$ can be presented by

$$\delta P(t) = A_P K_{inl}^P \left(1 - e^{-\frac{t}{\tau_P}}\right). \quad (56)$$

A.1.3 Model 1.3

Let us linearize Eqs. (6), (9) and (10):

$$\frac{LA}{\gamma RT} \frac{d(\Delta P)}{dt} = \Delta W_{inl} - \Delta W_{out}; \quad (57)$$

$$\frac{d(\Delta W_{inl})}{dt} = \frac{2A}{L}\left[\Delta P_{inl} - \Delta P - \frac{\xi RT}{2A^2}\left(\frac{2W_0}{P_{inl\ 0}}\Delta W_{inl} - \frac{W_0^2}{P_{inl\ 0}^2}\Delta P_{inl}\right)\right]; \quad (58)$$

$$\frac{d(\Delta W_{out})}{dt} = \frac{2A}{L}\left[\Delta P - \Delta P_{out} - \frac{\xi RT}{2A^2}\left(\frac{2W_0}{P_0}\Delta W_{out} - \frac{W_0^2}{P_0^2}\Delta P\right)\right]. \quad (59)$$

Next, we differentiate the equation for the pressure and substitute the derivatives of airflow. Then, considering $\frac{P_0}{P_{inl\ 0}} \Delta W_{inl} - \Delta W_{out} \approx \frac{LA}{\gamma RT} \frac{d(\Delta P)}{dt}$, we get

$$\frac{L^2}{2\gamma RT} \frac{d^2(\Delta P)}{dt^2} = \left(\frac{1}{2} + \frac{\xi RTW_0^2}{4A^2 P_{inl\ 0}^2}\right)\Delta P_{inl} + \frac{1}{2}\Delta P_{out} - \left(1 + \frac{\xi RTW_0^2}{4A^2 p_0^2}\right)\Delta P - \frac{\xi RTW_0}{2A^2 p_0} \frac{LA}{2\gamma RT} \frac{d(\Delta P)}{dt}. \quad (60)$$

In a relative deviations format

$$\frac{\tau_0^2}{2 + 0.5\xi\gamma M_0^2} \frac{d^2(\delta P)}{dt^2} + \frac{\tau_P}{1 + 0.25\xi\gamma M_0^2} \frac{d(\delta P)}{dt} + \delta P = \frac{2 + \xi\gamma M_0^2}{4 + \xi\gamma M_0^2}\delta P_{inl} + \frac{2}{4 + \xi\gamma M_0^2} \frac{P_{out\ 0}}{P_0} \delta P_{out}. \quad (61)$$

The transient process, which is initiated by the inlet pressure perturbation $\delta P_{inl} = A_P$, is expressed as

$$P(t) = A_P\left[1 - e^{-\frac{\xi}{2\tau}t}\left(\cos\omega t + \frac{\xi}{2\sqrt{2}} \sin\omega t\right)\right], \quad (62)$$

where $\omega \approx \frac{\sqrt{2}}{\tau_0}$.

A.1.4 Model 2.1

The linearization of Eq. (17) outputs the equation in the absolute deviations

$$\frac{PLA}{\gamma RG}\frac{d(\Delta T)}{dt} = -T_{inl\ 0}\Delta T + T_0 \Delta T_{inl}, \tag{63}$$

which in the relative deviations has the following form:

$$\tau_T \frac{d(\delta T)}{dt} + \delta T = \delta T_{inl}, \tag{64}$$

where $\tau_T = \frac{PAL}{\gamma RG_0 T_0} = \frac{\tau}{\gamma}$ is a time constant.

The transient process, which is initiated by the inlet temperature perturbation $\delta T_{inl} = A_T$, is described as

$$\delta T(t) = A_T\left(1 - e^{-\frac{t}{\tau_T}}\right). \tag{65}$$

A.1.5 Model 2.2

Let us transform Eq. (18) and linearize it:

$$\frac{L}{\gamma}\sqrt{\frac{\xi}{2R}}\frac{dP}{dt} = T\left[\sqrt{\frac{P_{inl}(P_{inl} - P)}{T_{inl}}} - \sqrt{\frac{P(P - P_{out})}{T}}\right]; \tag{66}$$

$$\frac{L}{\gamma}\sqrt{\frac{\xi}{2R}}\frac{d(\Delta P)}{dt} = \frac{W_0}{2A}\sqrt{\frac{\xi R}{2}}(\Delta T_{inl} - \Delta T) + \frac{A}{2W_0}\sqrt{\frac{2}{\xi R}}[(2P_{inl\ 0} - P_0)\Delta P_{inl} + P_0\Delta P_{out} - (P_{inl\ 0} + 2P_0 - P_{out\ 0})\Delta P]. \tag{67}$$

Next, we transform the coefficients and change the equation to the relative deviations:

$$\tau_P' \frac{d(\delta P)}{dt} + \delta P = K_T^P(\delta T - \delta T_{inl}) + K_{inl}^P \delta P_{inl} + K_{out}^P \delta P_{out}, \tag{68}$$

where $\tau_P' = 2\tau_P \frac{P_0}{P_{inl\ 0} + 2P_0 - P_{out\ 0}}$; $K_T^P = \frac{1}{2}\xi\gamma M_0^2 \frac{P_0}{P_{inl\ 0} + 2P_0 - P_{out\ 0}}$.

In a similar manner we transform Eq. (19):

$$\frac{2\tau}{\gamma + 1}\frac{d(\delta T)}{dt} + \delta T = \delta T_{inl} + K_{inl}^T \delta P_{inl} + K_{out}^T \delta P_{out} - K_p^T \delta P, \tag{69}$$

where $K_{inl}^T = \frac{2(\gamma-1)}{\gamma(\gamma+1)\xi M_0^2}\frac{2P_{inl\ 0} - P_0}{P_0}\frac{P_{inl\ 0}}{P_0}$; $K_{out}^T = \frac{2(\gamma-1)}{\gamma(\gamma+1)\xi M_0^2}\frac{P_{out\ 0}}{P_0}$;

$$K_p^T = \frac{2(\gamma - 1)}{\gamma(\gamma + 1)\xi M_0^2}\frac{P_{inl\ 0} + 2P_0 - P_{out\ 0}}{P_0}.$$

Having combined (68) and (69), we will get the differential equations for pressure and temperature:

$$\frac{\tau\tau_P}{\gamma}\frac{d^2(\delta P)}{dt^2}+\left(\frac{\tau}{\gamma}+\frac{\gamma+1}{2\gamma}\tau_P\right)\frac{d(\delta P)}{dt}+\delta P=\frac{\tau}{2\gamma}\left(\frac{d\delta P_{inl}}{dt}+\frac{d\delta P_{out}}{dt}\right)+\frac{1}{2}(\delta P_{inl}+\delta P_{out})-\frac{\tau}{4}\xi M_0^2\frac{d(\delta T_{inl})}{dt} \quad (70)$$

$$\frac{\tau\tau_P}{\gamma}\frac{d^2(\delta T)}{dt^2}+\left(\frac{\tau}{\gamma}+\frac{\gamma+1}{2\gamma}\tau_P\right)\frac{d(\delta T)}{dt}+\delta T=\frac{\gamma+1}{2\gamma}\tau_P\frac{d\delta T_{inl}}{dt}+\delta T_{inl}-\frac{\gamma-1}{\gamma}\frac{\tau_P}{\xi\gamma M_0^2}\left(\frac{d(\delta P_{inl})}{dt}+\frac{d(\delta P_{out})}{dt}\right). \quad (71)$$

The transient that is initiated by the pressure perturbation A_P is described as

$$\delta P(t)=\frac{A_P}{2\tau_P}\left(Ae^{at}+Be^{bt}+\tau_P\right), \quad (72)$$

where $A=\frac{a+d}{a(a-b)}$; $B=\frac{b+d}{b(b-a)}$; $d=\frac{\gamma}{\tau}$;

$$a=\frac{1}{\tau\tau_P}\left[-2\tau-(\gamma+1)\tau_P-\sqrt{(2\tau+(\gamma+1)\tau_P)^2-16k\tau\tau_P}\right];$$

$$b=\frac{1}{\tau\tau_P}\left[-2\tau-(\gamma+1)\tau_P+\sqrt{(2\tau+(\gamma+1)\tau_P)^2-16\gamma\tau\tau_P}\right];$$

$$\delta T(t)=\frac{A_P(\gamma-1)}{\xi\gamma\tau M_0^2}\left(A_1e^{at}+B_1e^{bt}\right), \quad (73)$$

where $A_1=\frac{a}{a(a-b)}$; $B_1=\frac{b}{b(b-a)}$.

The transient state, which is initiated by the temperature perturbation $\delta T_{inl}=A_T$, is described as

$$\delta P(t)=-\frac{A_T\xi\gamma M_0^2}{4\tau_P}\left(A_1e^{at}+B_1e^{bt}\right); \quad (74)$$

$$\delta T(t)=\frac{A_T(\gamma+1)}{2\tau}\left(A_2e^{at}+B_2e^{bt}+\frac{2\tau}{\gamma+1}\right), \quad (75)$$

where $A_2=\frac{a+d_1}{a(a-b)}$; $B_1=\frac{b+d_1}{b(b-a)}$; $d_1=\frac{2\gamma}{\tau_P(\gamma+1)}$.

A.1.6 Model 3.1

The linearized Eqs. (6), (7), (20), and (21) in the relative deviations format are

$$\frac{\tau}{\gamma}\frac{d(\delta P)}{dt}=\delta W_{inl}-\delta W_{out}+\delta T_{inl}-\delta T; \quad (76)$$

$$\frac{\tau}{\gamma}\frac{d(\delta T)}{dt}=\delta T+\delta T_{inl}+\frac{\gamma-1}{\gamma}(\delta W_{inl}-\delta W_{out}); \quad (77)$$

$$\frac{1}{2}\gamma\tau_0 M_0(\delta W_{inl}-\delta W_{out})=\delta P_{inl}+\delta P_{out}-2\delta P. \quad (78)$$

Let us transform Eqs. (76)–(78) to get the differential equations for the pressure and the temperature:

$$\frac{1}{4}\tau^2\tau_0 M_0\frac{d^3(\delta P)}{dt^3}+\frac{1}{4}\gamma\tau\tau_0 M_0\frac{d^2(\delta P)}{dt^2}+\tau\frac{d(\delta P)}{dt}+\delta P=$$
$$=\frac{1}{4}\gamma\tau\tau_0 M_0\frac{d^2T_{inl}}{dt^2}+\frac{1}{2}\tau\left(\frac{d(\delta P_{inl})}{dt}+\frac{d(\delta P_{out})}{dt}\right)+\frac{1}{2}(\delta P_{inl}+\delta P_{out}). \tag{79}$$

$$\frac{1}{4}\tau^2\tau_0 M_0\frac{d^3(\delta T)}{dt^3}+\frac{1}{4}\gamma\tau\tau_0 M_0\frac{d^2(\delta T)}{dt^2}+$$
$$+\tau\frac{d(\delta T)}{dt}+\delta T=\frac{1}{4}\gamma\tau\tau_0 M_0\frac{d^2T_{inl}}{dt^2}+\delta T_{inl}+\frac{\gamma-1}{2\gamma}\tau\left(\frac{d(\delta P_{inl})}{dt}+\frac{d(\delta P_{out})}{dt}\right). \tag{80}$$

The transient that is initiated by the temperature perturbation $\delta T_{inl}= A_T$ is described as

$$\delta T(t)=A_T\frac{\gamma}{\tau}\left(A_3e^{\alpha t}\sin(\omega t+\beta_1)+B_3e^{s_1t}+K\right), \tag{81}$$

where α and ω are expressed by (38) and (39); $A_3=\frac{1}{\omega}\sqrt{\frac{\left(\alpha^2-\omega^2+\frac{4}{\gamma\tau_0^2}\right)^2+(2\alpha\omega)^2}{(\alpha^2+\omega^2)\left[(\alpha-s_1)^2+\omega^2\right]}}$;
$B_3=\frac{s_1^2+\frac{4}{\gamma\tau_0^2}}{s_1\left[(s_1-\alpha)^2+\omega^2\right]}$; $K=-\frac{4}{\gamma\tau_0^2 s_1(\alpha^2+\omega^2)}$; $\beta=\operatorname{arctg}\frac{2\alpha\omega}{\alpha^2-\omega^2+\frac{4}{\gamma\tau_0^2}}-\operatorname{arctg}\frac{\omega}{\alpha-s_1}-\operatorname{arctg}\frac{\omega}{\alpha}$;

$$\delta P(t)=A_T\frac{\gamma}{\tau}\left(A_4e^{\alpha t}\sin(\omega t+\beta)+B_4e^{s_1t}\right), \tag{82}$$

where $A_4=\frac{1}{\omega}\sqrt{\frac{(\alpha^2-\omega^2)^2+(2\alpha\omega)^2}{(\alpha^2+\omega^2)\left[(\alpha-s_1)^2+\omega^2\right]}}$; $B_4=\frac{s_1^2}{s_1\left[(s_1-\alpha)^2+\omega^2\right]}$.
The transient that is initiated by the pressure perturbation A_P is described as

$$\delta T(t)=A_T\frac{2(\gamma-1)}{\gamma\tau_0^2}\left(A_5e^{\alpha t}\sin(\omega t+\beta_1)+B_5e^{s_1t}\right), \tag{83}$$

where $A_5=\frac{1}{\omega}\sqrt{\frac{1}{(\alpha-s_1)^2+\omega^2}}$; $B_5=\frac{1}{(s_1-\alpha)^2+\omega^2}$; $\beta_1=-\operatorname{arctg}\frac{\omega}{\alpha-s_1}$;

$$\delta P(t)=A_T\frac{2}{\tau_0^2}\left(A_6e^{\alpha t}\sin(\omega t+\beta_2)+B_6e^{s_1t}+K_2\right), \tag{84}$$

where $A_6=\frac{1}{\omega}\sqrt{\frac{\left(\frac{\tau_0^2}{2\tau}+\alpha\right)^2+\omega^2}{(\alpha^2+\omega^2)\left[(\alpha-s_1)^2+\omega^2\right]}}$; $B_6=\frac{s_1+\frac{\tau_0^2}{2\tau}}{s_1\left[(s_1-\alpha)^2+\omega^2\right]}$; $K_2=-\frac{\tau_0^2}{2\tau s_1(\alpha^2+\omega^2)}$;
$\beta_2=-\operatorname{arctg}\frac{\omega}{\alpha+\frac{\tau_0^2}{2\tau}}-\operatorname{arctg}\frac{\omega}{\alpha-s_1}-\operatorname{arctg}\frac{\omega}{\alpha}$.

A.1.7 Model 3.2

The model consists of Eqs. (6), (7), (9) and (10). Linearized Eqs. (6) and (7) are of the format presented in Eqs. (76) and (78). As a result of the linearization, we get the missing difference $\frac{dW_{inl}}{dt}-\frac{dW_{out}}{dt}$:

$$\frac{dW_{inl}}{dt} - \frac{dW_{out}}{dt} = \frac{2}{\gamma\tau_0 M_0}\{\left(\frac{P_{inl\,0}}{P_0} + \xi\gamma M_0^2\frac{P_0}{P_{inl\,0}}\right)\delta P_{inl} + \frac{P_{out\,0}}{P_0}\delta P_{out} - (2 + \xi\gamma M_0^2) + \delta P$$
$$-\xi\gamma M_0^2[2(\delta W_{inl} - \delta W_{out}) + \delta T_{inl} - \delta T]\}. \tag{85}$$

On the other hand, from Eq. (77) we get

$$\frac{dW_{inl}}{dt} - \frac{dW_{out}}{dt} = \frac{\gamma}{\gamma - 1}\left(\frac{\tau}{\gamma}\frac{d^2(\delta T)}{dt^2} + \frac{d(\delta T)}{dt} - \frac{d(\delta T_{inl})}{dt}\right). \tag{86}$$

Let us determine $\frac{d(\delta P)}{dt}$ from (76) and (78):

$$\frac{d(\delta P)}{dt} = \frac{\gamma}{\gamma - 1}\left[\frac{d(\delta T)}{dt} + \frac{1}{\tau}(\delta T - \delta T_{inl})\right]. \tag{87}$$

Having equalized the right sides of Eqs. (85) and (86), derived the obtained equation, and substituted the derivative (87) in it, we get a differential equation for the temperature in the volume:

$$\frac{\tau\tau_0^2}{2C}\frac{d^3(\delta T)}{dt^3} + \frac{\tau_0^2}{C}(0.5\gamma + 2\xi)\frac{d^2(\delta T)}{dt^2} + \frac{\tau}{C}\left[2 + (2\gamma + 1)\xi M_0^2\right]\frac{d(\delta T)}{dt} + \delta T$$
$$= \frac{\gamma\tau_0^2}{2C}\frac{d^2(\delta T_{inl})}{dt^2} + \frac{(\gamma + 1)\xi\tau_0 M_0}{C}\frac{d(\delta T_{inl})}{dt} + \delta T_{inl}$$
$$+ \frac{\tau(\gamma - 1)}{C}\left[\left(\frac{P_{inl\,0}}{P_0} + \xi\gamma M_0^2\right)\frac{d(\delta P_{inl})}{dt} + \frac{P_{out\,0}}{P_0}\frac{d(\delta P_{out})}{dt}\right] \tag{88}$$

where $C = 2 + \xi\gamma M_0^2$.

Let us use the Laplace transform and transfer functions $W_T^P(s)$, $W_{P_{inl}}^T(s)$, $W_{out}^T(s)$, and $W_{T_{inl}}^T(s)$ to obtain the equation for the pressure. The transfer functions we will get from Eqs. (87) and (88). The final result is

$$\frac{\tau\tau_0^2}{2C}\frac{d^3(\delta P)}{dt^3} + \frac{\tau_0^2}{C}(0.5\gamma + 2\xi)\frac{d^2(\delta P)}{dt^2} + \frac{\tau}{C}\left[2 + (2\gamma + 1)\xi M_0^2\right]\frac{d(\delta P)}{dt} + \delta P =$$
$$= \frac{\gamma\tau_0^2}{2C}\frac{d^2(\delta T_{inl})}{dt^2} + \frac{\tau_0\xi\gamma M_0}{C}\frac{d(\delta T_{inl})}{dt} + \frac{\tau}{C}\left[\left(\frac{P_{inl\,0}}{P_0} + \xi\gamma M_0^2\right)\frac{d(\delta P_{inl})}{dt} + \frac{P_{out\,0}}{P_0}\frac{d(\delta P_{out})}{dt}\right] +$$
$$+ \frac{1}{C}\left[\left(\frac{P_{inl\,0}}{P_0} + \xi\gamma M_0^2\right)\delta P_{inl} + \frac{P_{out\,0}}{P_0}\delta P_{out}\right]. \tag{89}$$

The transient state, which is initiated by the temperature perturbation δT_{inl}, (the magnitude of the step is A_T) is described as

$$\delta T(t) = A_T\frac{\gamma}{\tau}\left(\begin{array}{l} A_6 \cdot 1(t) - (A_6 + B_6)e^{s_1 t} + \\ + \dfrac{B_6\sqrt{\left(\frac{K_3}{B_6} - \alpha\right)^2 + \omega^2}}{\omega}e^{\alpha t} \cdot \sin(\omega t + \beta_3) \end{array}\right), \tag{90}$$

where α and ω are expressed by (48) and (39): $A_6 = -\frac{b_1}{s_1(\alpha^2+\omega^2)}$; $B_6 = \frac{(2\alpha s_1 - s_1^2)A_6 - s_1 - a_1}{\alpha^2+\omega^2 - s_1(2\alpha - s_1)}$; $K_3 = 1 + s_1 A_6 - (2\alpha - s_1)B_6$; $\beta_3 = \mathrm{arctg}\frac{\omega}{\frac{K_3}{B_6}-\alpha}$; $a_1 = \frac{2(\gamma+1)\xi M_0}{\gamma\tau_0}$; $b_1 = \frac{2C}{\gamma\tau_0^2}$;

$$\delta P(t) = A_T\frac{\gamma}{\tau}\left(A_7 e^{s_1 t} + \frac{B_7\sqrt{\left(\frac{K_4}{B_7}-\alpha\right)^2+\omega^2}}{\omega}e^{\alpha t}\cdot\sin(\omega t+\beta_4)\right), \tag{91}$$

where $A_7 = -B_7$; $B_7 = -\frac{a_2+s_1}{\alpha^2+\omega^2-s_1(2\alpha-s_1)}$; $K_4 = \frac{(\alpha^2+\omega^2)A_7 - a_2}{s_1}$; $\beta_4 = \mathrm{arctg}\frac{\omega}{\frac{D_2}{C_2}-\alpha}$; $a_2 = \frac{2\xi\gamma M_0}{\tau_0\gamma}$.

The transient that is caused by the pressure perturbation A_P is described as

$$\delta T(t) = A_P\frac{2(\gamma-1)}{\tau_0^2}\left(\frac{P_{\mathrm{inl}\,0}}{P_0}+\xi\gamma M_0^2\right)\left(A_8 e^{s_1 t} + \frac{B_8\sqrt{\left(\frac{K_5}{B_8}-\alpha\right)^2+\omega^2}}{\omega}e^{\alpha t}\cdot\sin(\omega t+\beta_5)\right), \tag{92}$$

where $A_8 = -B_8$; $B_8 = -\frac{1}{\alpha^2+\omega^2-s_1(2\alpha-s_1)}$; $K_5 = \frac{(\alpha^2+\omega^2)A_8 - 1}{s_1}$; $\beta_1 = \mathrm{arctg}\frac{\omega}{\frac{K_5}{B_8}-\alpha}$.

$$\delta P(t) = A_P\frac{2}{\tau_0^2}\left(\frac{P_{\mathrm{inl}\,0}}{P_0}+\xi\gamma M_0^2\right)\left(A_9\cdot 1(t) - (A_9+B_9)e^{s_1 t} + \frac{B_9\sqrt{\left(\frac{K_6}{B_9}-\alpha\right)^2+\omega^2}}{\omega}e^{\alpha t}\cdot\sin(\omega t+\beta_6)\right), \tag{93}$$

where $A_9 = -\frac{b_3}{s_1(\alpha^2+\omega^2)}$; $B_9 = \frac{(2\alpha s_1 - s_1^2)A_9 - 1}{\alpha^2+\omega^2-s_1(2\alpha-s_1)}$; $K_6 = s_1 A_9 - (2\alpha - s_1)B_9$; $\beta_6 = \mathrm{arctg}\frac{\omega}{\frac{K_6}{B_9}-\alpha}$; $b_3 = \frac{1}{\tau}$.

Author details

Sergiy Yepifanov and Roman Zelenskyi*
National Aerospace University "Kharkiv Aviation Institute", Kharkiv, Ukraine

*Address all correspondence to: aedlab@gmail.com

References

[1] Jaw LC, Mattingly JD. Aircraft Engine Controls: Design, System Analysis, and Health Monitoring. Reston, USA: AIAA; 2009. 294p

[2] Fawke AJ, Saravanamuttoo HIH. Digital Computer Methods for Prediction of Gas Turbine Dynamic Response. Report No. 710550. England: Society of Automotive Engineers, Station, Newcastle; 1971

[3] Fawke AJ. Digital computer simulation of gas turbine dynamic behavior [thesis]. Bristol, England: University of Bristol; 1970

[4] Glikman BF. Mathematical Models of Pneumatic Hydraulic Systems. Russia, Nauka: Moscow; 1986. 368p

[5] Gas Turb. Available from: http://www.gasturb.de

[6] Zhao YS, Hu J, Wu TY, Chen JJ. Steady-state and transient performances simulation of large civil aircraft auxiliary power unit. ISABE Paper ISABE-2011-1323; 2011. 10p

[7] Ghigliazza F, Traverso A, Pascenti M, Massardo AF. Micro gas turbine real-time modeling: Test rig verification. ASME Paper GT2009-59124; 2009. 8p

[8] Davison CR, Birk AM. Comparison of transient modeling techniques for a micro turbine engine. ASME Paper GT2006-91088; 2006. 10p

[9] Koçer G, Uzol O, Yavrucuk İ. Simulation of the transient response of a helicopter turboshaft engine to hot-gas ingestion. ASME Paper GT2008-51164; 2008. 6p

[10] Gang Y, Jianguo S, Xianghua H, Wei-Sheng S. A Non-Iterative Method of Aero-Engine Modeling Using Complementary Variables. Journal of Aerospace Power. Beijing, China: Chinese Society of Aeronautics and Astronautics; 2003

[11] Kong X, Wang X, Tan D, He A. A non-iterative aero engine model based on volume effect. AIAA Paper AIAA2011-6623; 2011

[12] Yepifanov SV, Zelenskyi RL. The simulation of the pneumatic volume dynamics as a part of transients in the gas path of gas turbine engines. Aerospace Technique and Technology. 2007;**10**(46):49-54

[13] Shi Y, Tu Q, Jiang P, Zheng H, Cai Y. Investigation of the compressibility effects on engine transient performance. ASME Paper GT2015-42889; 2015. 8p

[14] Culmone MV, Garcia-Rosa N, Carbonneau X. Sensitivity analysis and experimental validation of transient performance predictions for a short-range turbofan. ASME Paper GT2016-57257; 2016. 10p

[15] Wang C, Li Y-G, Yang B-Y. Transient performance simulation of aircraft engine integrated with fuel and control systems. ISABE Paper ISABE-2015-20103; 2015. 11p

[16] Kopasakis G, Connolly JW, Paxson DE, Ma P. Volume dynamics propulsion system modeling for supersonics vehicle research. ASME Paper GT2008-50524; 2008. 10p

[17] Henke M, Monz T, Aigner M. Introduction of a new numerical simulation tool to analyze micro gas turbine cycle dynamics. ASME Paper GT2016-56335; 2016. 11p

[18] Wang C, Li YG. Hydraulic fuel system simulation using Newton-Raphson method and its integration

with a gas turbine performance model. ASME Paper GT2017-63881; 2017. 11p

[19] Mohajer A, Abbasi E. Development of compression system dynamic simulation code for testing and designing of anti-surge control system. ASME Paper GT2017-63212; 2017. 8p

[20] Shevyakov AA. Automatic Control Systems of the Air-Breathing Power Plants. Moscow: Mechanical Engineering; 1992. 432p

[21] Dobryansky GV, Martyanova TS. Dynamics of Aircraft Engines. Mechanical Engineering: Moscow; 1989. 240p

[22] Kurosaki M, Sasamoto M, Asaka K, Nakamura K, Kakiuchi D. An efficient transient simulation method for a gas turbine volume dynamics model. ASME Paper GT2018-75353; 2018. 11p

[23] Gurevich O. Aircraft Engines Automatic Control Systems. Encyclopedic Guide. Moscow: Torus Press; 2011. 208p

[24] Chen CT. Introduction to Linear System Theory. Philadelphia: Holt, Rinehart and Winston; 1970

[25] Bryson AE. Control of a Spacecraft and Aircraft. Princeton, NJ: Princeton University Press; 1994

[26] Korn GA, Korn TM. Mathematical Handbook for Scientists and Engineers. Definitions, theorems and Formulas for Reference and Review. New York/Toronto/London: McGraw Hill Book Company, Inc.; 1961. pp. 47-48

Chapter 4

A New Approach for Model Developing to Estimate Unmeasured Parameters in an Engine Lifetime Monitoring System

Cristhian Maravilla and Sergiy Yepifanov

Abstract

Monitoring systems to predict the remaining lifetime of gas turbine engines are a major field of investigation, in particular, the monitoring systems that allow an on-line prediction. This chapter introduces and analyzes a new approach to develop mathematical models to estimate unmeasured parameters in an engine lifetime monitoring system; these models in contrast to previously developed models allow an on-line monitoring of unmeasured parameters, which are necessary to perform an on-line lifetime prediction. The blade of a high-pressure turbine (HPT) of a two-spool free turbine power plant is the test case. Several candidate models are developed for each unmeasured parameter; the best models are selected by their accuracy and robustness using the instrumental and truncation error as criteria. Ten faulty engine conditions are considered to analyze the model robustness. Two methods for model developing are compared; the first method uses physics-based models (proposed in this chapter), and the second method develops the models using the similarity concept (reference methodology). The results of the comparison show that the physics-based models are more robust to engine faults and overall they deliver a significantly more accurate prediction of the engine lifetime.

Keywords: gas turbine, lifetime prediction, model developing, thermodynamic relations, unmeasured parameters

1. Introduction

Lifetime monitoring systems are an effective way to perform condition base maintenance of gas turbine engines [1–3]; this allows a better use of the available lifetime and improvement of the engine's reliability.

Several approaches exist to predict the remaining lifetime, such as neural networks [4–7], finite element analysis (FEA) [8–10], and statistical methods [11]; however, in order to significantly enhance the accuracy of the lifetime prediction, it is necessary to estimate the lifetime in real time (on-line prediction) using actual conditions [12, 13]. All of the previously cited approaches require a large amount of computing resources, making them not suitable for an on-line application; another limitation is that none of the cited approaches take into consideration the existing engine-to-engine differences and the performance deterioration.

Oleynik proposes in a study [14] an approach to perform on-line lifetime prediction of engine component condition (**Figure 1**); he proposes to use simple models to estimate the temperature and stresses at critical points. A major advantage of this approach is the use of actual engine operating and atmospheric conditions as input data. The methodology has been proven by practical application in monitoring of some Ukrainian engines.

As shown in **Figure 1**, one of the blocks performs the thermal condition (TC) monitoring, while the second block, the stress condition (SC) monitoring.

Inside the block of TC, it is necessary to set the thermal boundary conditions (gas temperature around the critical element and the heat transfer coefficient between the gas and metal), which are not measured parameters; the author [14] proposed a methodology to develop mathematical models based on the theory of similarity to estimate these boundary conditions.

In this chapter a new approach to develop physics-based models to estimate the unmeasured parameters is proposed and analyzed. This new approach emphasizes the accuracy of the models regardless of the engine-to-engine differences, taking into consideration a healthy and faulty engine condition.

With the help of the thermodynamic model of the engine chosen as test case, all the necessary data for model developing and validation is simulated; the simulation of the engine's component degradation is widely used in gas turbine monitoring and diagnostics [15, 16].

A comparison between the physics-based models and the reference method based on the theory of similarity proposed by [14] is conducted to validate the accuracy of the methods.

Finally, the influence of the accuracy in the prediction of the thermal boundary conditions on the errors of the engine lifetime prediction is evaluated.

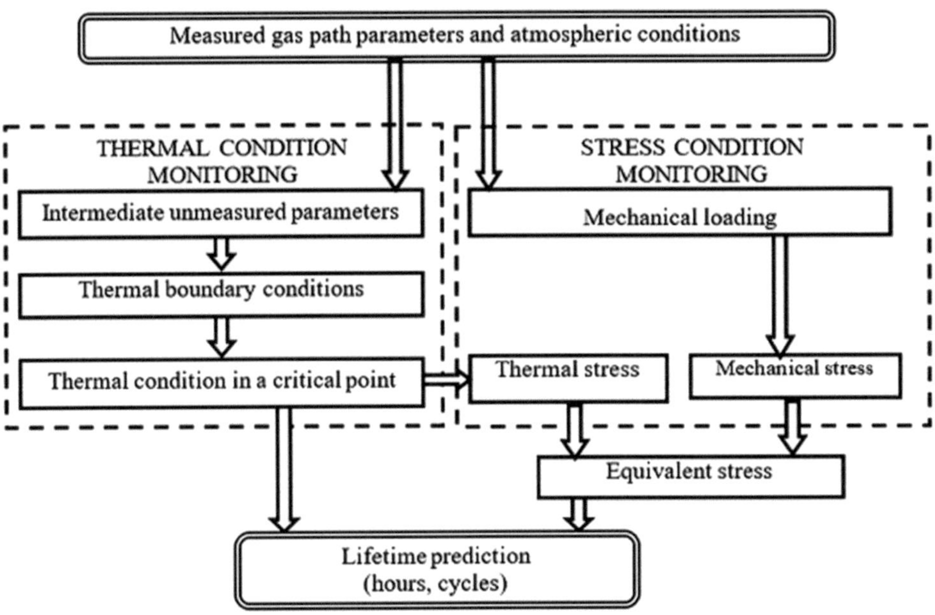

Figure 1.
Scheme for on-line lifetime prediction.

2. Test case

It is well-known that the thermomechanical stresses in gas turbine hot elements are very high, particularly in the turbine blade having a significant effect on the safety and economics of the fleet operation [17]. For this reason, a turbine blade with three cooling channels has been chosen as the test case; the blade is mounted in the first stage of the high-pressure turbine (HPT) of a two-spool free turbine engine.

In **Table 1** the seven measured gas path parameters of the engine chosen as test case are listed.

Experience has shown that it is sufficient to consider a two-dimensional analysis of the mid-span section of the blade in order to save computing time; however, the same methodological approach can be applied for a three-dimensional geometry.

The finite element model of the mid-span turbine blade section was built with the help of FEA software. After setting all the necessary boundary conditions, the distribution of the temperature and stresses was obtained. The critical points with the smallest safety factor were found as well; as shown in **Figure 2**, the critical points correspond to the numbers 101, 102, 103, and 69.

Three critical points are located at the leading edge of the blade and one critical point at the trailing edge. For an easier analysis, the four critical points are

Designation	Gas path parameter
G_f	Fuel consumption
T_C^*	Compressor discharge temperature
p_C^*	Compressor discharge pressure
T_{HPT}^*	HPT discharge temperature
p_{HPT}^*	HPT discharge pressure
T_{LPT}^*	Low-pressure turbine (LPT) discharge temperature
n_{HP}	Rotational speed of the high-pressure (HP) rotor

Table 1.
List of gas path measured parameters of the engine selected as test case.

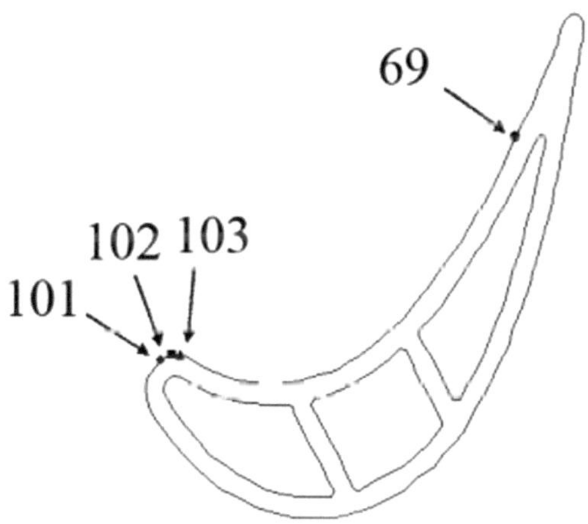

Figure 2.
Critical points at the turbine blade mid-span section.

organized into two groups: group a (GA) contains the critical points 101, 102, and 103, and group b (GB) contains the critical point number 69.

The thermodynamic model of the engine chosen as test case [18] is used to generate all the data for model developing and validation.

3. Engine lifetime monitoring system

A brief description of the lifetime monitoring system proposed by the author [14] is presented to have a better understanding. As shown in **Figure 1**, the block to perform the TC contains the sub-block named "thermal boundary conditions" (unmeasured parameters). The new approach for model developing proposed in this chapter focuses all the efforts in improving the accuracy in this sub-block, as improving the efficiency in the prediction of the thermal boundary conditions directly affects the accuracy of the whole monitoring system.

In the following subsection, the simple mathematical models used to estimate the blade temperature and the thermal stress at the critical points are explained.

3.1 Turbine blade thermal and stress monitoring models

In [19] an analysis of the models proposed by [14] to estimate the blade temperature and stress applied to the same turbine blade and critical points was conducted. As a result, it was concluded that the best model to calculate the blade temperature at the critical points is:

$$t_{cr} = T^*_{S1} + \Theta(k_\alpha) \cdot \left(T^*_{S2} - T^*_{S1}\right) \tag{1}$$

Here t_{cr} is the blade temperature at the critical point, T^*_{S1} is the cooling temperature, T^*_{S2} is the heating temperature, Θ is a dimensionless parameter, and k_α is the relation of the heat transfer coefficients at current and reference engine operating modes:

$$k_\alpha = {}^{\alpha_i}\!/_{\alpha_{ref}} \tag{2}$$

Here α_i and α_{ref} are the heat transfer coefficients at actual and reference engine working operating modes, respectively.

The model to estimate the thermal stress at critical points is:

$$\sigma_t = \overline{S}(k_\alpha) \cdot n^2 \tag{3}$$

Here σ_t is the thermal stress at the critical point, $\overline{S}$ is a dimensionless parameter, and n is the rotational speed.

The dimensionless parameters Θ in Eq. (1) and $\overline{S}$ in Eq. (3) are calculated as dependence of k_α using a polynomial with gas path measured parameters as arguments.

The parameters T^*_{S1}, T^*_{S2}, and k_α are unmeasured parameters, which further will be referred as thermal boundary conditions; since this unmeasured parameters are the input data in Eqs. (1) and (3), it is necessary to develop mathematical models for their prediction. In the following section, the model developing methodology is explained.

4. Model developing methodology

As mentioned in the introduction, a new approach for model developing is proposed. Since the models will be used in an on-line monitoring system, it is necessary to meet the following requirements:

- The models must be developed on physics-based relations, such as thermodynamic and kinematic relations and others.
- All models must use the gas path measured parameters as input data.
- The structure of the models must be simple.
- The measuring error of the gas path parameters, which are the input data of the models, must be taken into account.
- The models must have high robustness to the influence of engine's component deterioration.

Taking into account the main requirements, the following general dependence for any unmeasured parameter is proposed:

$$z = f\left(Y, W, T_H^*, P_H^*\right) \tag{4}$$

Here z is the vector of unmeasured parameters to be predicted; Y is the vector of gas path measured parameters; W is the vector of the intermediate unmeasured parameters which describe the thermodynamic properties of the fluid, such as efficiencies and pressure loss factors; and T_H^*, P_H^* are the ambient conditions.

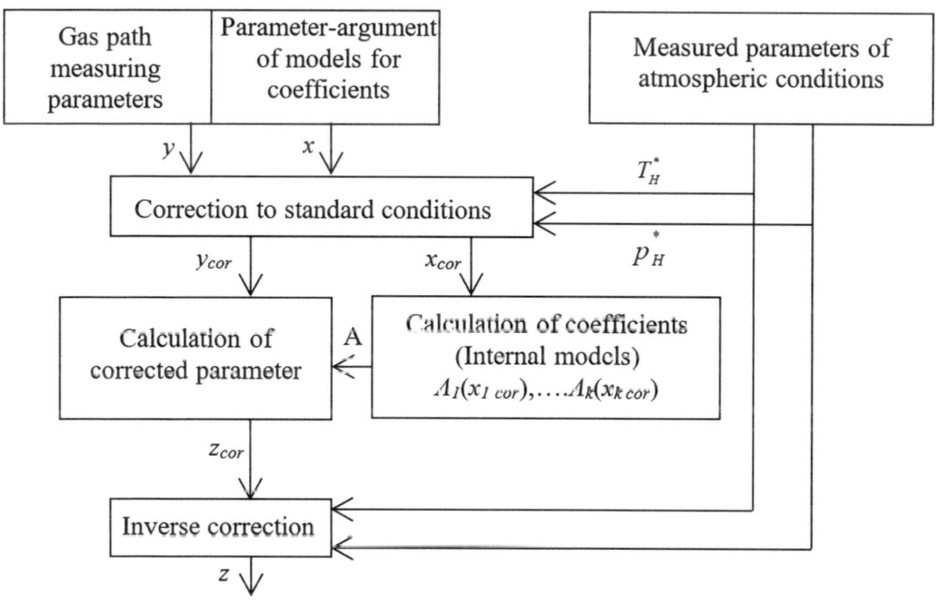

Figure 3.
Algorithm to estimate the unmeasured parameters.

As shown in Eq. (4), such dependence is not possible to be used in an on-line monitoring system since it includes the vector W, which is an unmeasured parameter. In order to solve this inconvenience, it is proposed to estimate each unmeasured parameter w as a function A (internal model) using one of the available gas path measured parameters as argument $w = A(x)$, where x is a gas path measured parameter.

Besides, all the parameters are corrected to standard atmosphere to take into account the influence of atmospheric conditions [20]. After rewriting Eq. (4), the general dependence for any unmeasured parameter is:

$$z_{cor} = f(Y_{cor}, A(x_{cor})) \tag{5}$$

Figure 3 shows the algorithm to estimate the unmeasured parameters using the previously presented model developing methodology.

4.1 Model verification

The best models are selected during the model verification process. The total mean square error (MSE) is the main criteria to select the best models; the MSE consists of two components, a truncation error and an instrumental error:

$$\sigma_{TT} = \sigma_{TR} + \sigma_{INS} \tag{6}$$

Here σ_{TT} is the total MSE, σ_{TR} is the truncation error, and σ_{INS} is the instrumental error.

4.1.1 Truncation error

The mean square truncation error is:

$$\sigma_{TR\,j} = \sqrt{\frac{\sum_{j=1}^{N_j} \left(z_{j\,i\,m} - z_{j\,i}\right)^2}{N_j}} \tag{7}$$

Here $z_{j\,i\,m}$ is the value of the unmeasured parameter calculated by the developed models, $z_{j\,i}$ is the true value obtained from the engine thermodynamic model, and N_j is the sample size; in other words, it is the number of engine operating points considered for analysis corresponding the engine health condition number j.

The average truncation error in percentage for any engine health conditions is obtained by:

$$\sigma_{TR\,m} = \frac{1}{n}\sum_{j=1}^{n} \sigma_{TR\,\mathrm{j}} \cdot 100\% \tag{8}$$

Here n is the number of engine health conditions.

4.1.2 Instrumental error

The MSE for the instrumental error for a healthy engine condition is calculated as follows:

$$\sigma_{INS} = \sqrt{\sum_{q=1}^{Q}\left(\frac{\partial z}{\partial y_q}\right)_i \cdot \sigma_{y\,q}^2 + \sum_{K=1}^{K}\left(\frac{\partial z}{\partial x_K}\right)_i \cdot \sigma_{x\,K}^2 + \left(\frac{\partial z}{\partial T_H^*}\right)_i \cdot \sigma_{T_H^*}^2 + \left(\frac{\partial z}{\partial p_H^*}\right)_i \cdot \sigma_{p_H^*}^2} \quad (9)$$

Here Q is the amount of gas path measured parameters, and K is the amount of internal models, which are part of the developed model to estimate the unmeasured parameter$z$.

The average instrumental error in percentage is obtained by:

$$\sigma_{INS\ m} = \frac{1}{N}\sum_{j=1}^{n} \sigma_{INS\ j} \cdot 100\% \quad (10)$$

Here N is the sample size—the number of engine operating points corresponding to a healthy engine condition.

4.1.3 Model robustness analysis

Engine health conditions in real life are different from engine to engine; therefore, it is necessary to take into account these shifts from the ideal engine. For the model robustness analysis, the truncation error is calculated for several engine health conditions. In [21] the most common engine faulty conditions of a two-spool free turbine engine were analyzed; as a result, 10 faulty engine conditions were selected.

In **Table 2** the 11 engine conditions considered for analysis are listed. The deteriorated engine condition represents a 3% shift from engine health condition.

Designation	Fault parameter	Deteriorated condition
C1	—	Healthy engine
C2	$\delta\eta_C$	Compressor efficiency decrease
C3	δG_C	Compressor airflow decrease
C4	$\delta\eta_{CC}$	Combustion chamber (CC) efficiency decrease
C5	$\delta\sigma_{CC}$	Decrease in the CC total pressure
C6	$\delta\eta_{HPT}$	HPT efficiency decrease
C7	δF_{NBHPT}	HPT nozzle box (NB) area increment
C8	$\delta\sigma_{HPT-LPT}$	Decrease in the total pressure between HPT and LPT duct
C9	$\delta\eta_{LPT}$	LPT efficiency decrease
C10	δF_{NBLPT}	LPT NB area increment
C11	δG_{ST}	Increment of the air bleed for gas pumping needs

Table 2.
Engine healthy and faulty condition considered for analysis.

5. Developing and verification of models for unmeasured parameters for the test case

As shown in SubSection 3.1, it is necessary to calculate the input data for Eqs. (1) and (3). For our particular test case, the selected thermal boundary conditions are shown in **Table 3**.

Cooling temperature	$T^*_{S1} = T^*_C$
Heating temperature	$T^*_{S2} = T^*_W$
Relation of heat transfer coefficients for hot gases	$k_{\alpha\ g} = {\alpha_{i\ g}}/{\alpha_{ref\ g}}$
Relation of heat transfer coefficients for cooling air	$k_{\alpha\ a} = {\alpha_{i\ a}}/{\alpha_{ref\ a}}$
Here T^*_C is the compressor discharge temperature, and T^*_W is the gas temperature in relative motion in front of the first turbine stage.	

Table 3.
Thermal boundary conditions for our test case.

It is necessary to develop alternative models for each one of the selected thermal boundary conditions. Although the compressor discharge temperature is a measured gas path parameter for our test case (**Table 1**), it is of particular interest to analyze the effect of its inclusion or exclusion from the list of gas path measured parameters in the overall lifetime prediction.

Since many alternative models were developed, only one example will be shown in order to save available text space; for more details consult [21].

5.1 Model developing

Let us consider the gas temperature at the inlet of the turbine T^*_g. As mentioned in Section 4, the models must be physics-based; therefore, we use the relation that describes the power balance between the high-pressure turbine and the high-pressure compressor [20] as the base equation for model developing:

$$N_T \cdot \eta_m = N_C \tag{11}$$

Here N_T is the turbine power, η_m is the mechanical efficiency, and N_C is the compressor power.

Let us take into account that:

$$N_T = G_g \cdot L_T = G_g \cdot C_{p\ g} \cdot \left(T^*_g - T^*_T\right) \tag{12}$$

$$N_C = G_a \cdot L_C = G_a \cdot C_{p\ a} \cdot \left(T^*_C - T^*_H\right) \tag{13}$$

Here G_g and G_a are the gas and air flow consumptions accordingly, and $C_{p\ g}$ and $C_{p\ a}$ are the specific heat values at constant pressure for gas and air accordingly.

Solving Eq. (11) for T^*_g, we obtain:

$$T^*_g = \left(\frac{C_{p\ a}}{C_{p\ g}} \cdot \frac{G_a}{G_g \cdot \eta_m}\right) \cdot \left(T^*_C - T^*_H\right) + T^*_{HPT} \tag{14}$$

All the unmeasured parameters in Eq. (14) will be described with the help of the internal model A_{TG1}:

$$A_{TG1} = \left(\frac{C_{p\ a}}{C_{p\ g}} \cdot \frac{G_a}{G_g \cdot \eta_m}\right) \tag{15}$$

Let us remember that all the internal models will be calculated as a polynomial functions using one of the available gas path measured parameters as argument.

Finally, after making the correction to standard atmospheric conditions of Eq. (14), we arrive at the following expression:

$$T^*_{g\ cor} = A_{TG1\ cor}(x) \cdot \left(T^*_{C\ cor} - T_0\right) + T^*_{HPT\ cor} \tag{16}$$

Here T_0 = 288.15 K is the value of standard atmospheric temperature.

Several models can be developed for the same unmeasured parameter based on different source equations (alternative models). **Figure 4** shows the general scheme for the developing of alternative models.

Using the scheme shown in **Figure 4**, a total of 31 models were developed for our test case (see reference [21]). A name is given to each developed model for easier identification, for example, the model shown in Eq. (16) is named MTG1. The names and arguments for each alternative developed model are shown in **Figure 5**.

5.2 Model verification

Let us consider the model MTG1 shown in Eq. (16), which includes the internal model $A_{TG1\ cor}(x)$. It is necessary to analyze which of the gas path measured parameters (**Table 1**) is best suited to be used as argument in the polynomial, as well as the degree of the polynomial that results in the most accurate prediction of T^*_g.

All of the necessary data for the analysis was obtained using the thermodynamic model of the engine selected as test case [18]. A total of 245 engine operating modes for a healthy engine condition were simulated; these operating modes describe the whole range of the engine operating conditions.

The generated data was randomly divided into two sets. The first set of 123 operating points is used as reference. The second set of 122 operation modes is for model validation. Using the data from the reference set, the coefficients for the polynomial functions were obtained using the least square method.

The polynomial degree was changed from 1 to 4 using each one of the gas path measured parameters (see **Table 1**) as argument in the polynomial.

Once all the polynomial coefficients describing the internal model $A_{TG1\ cor}(x)$ were obtained, the value of T^*_g was calculated. According to the scheme shown in

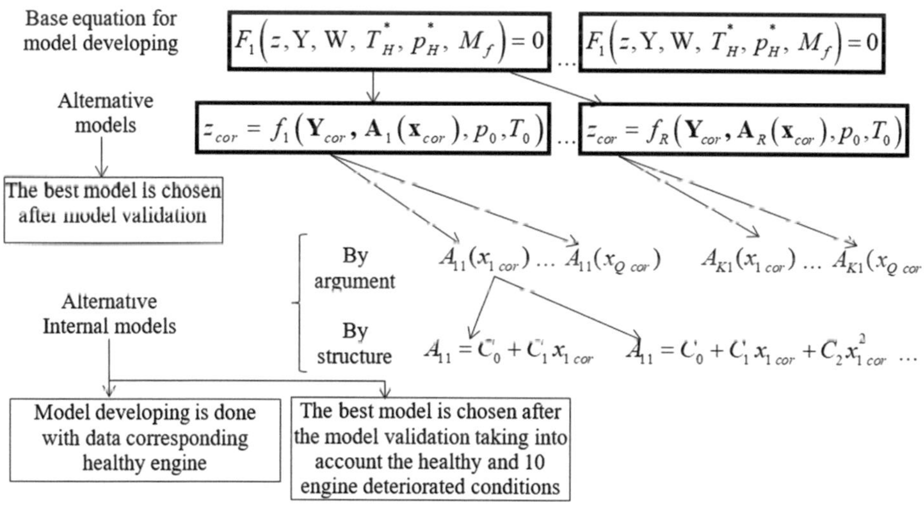

Figure 4.
Scheme for the developing the alternative models.

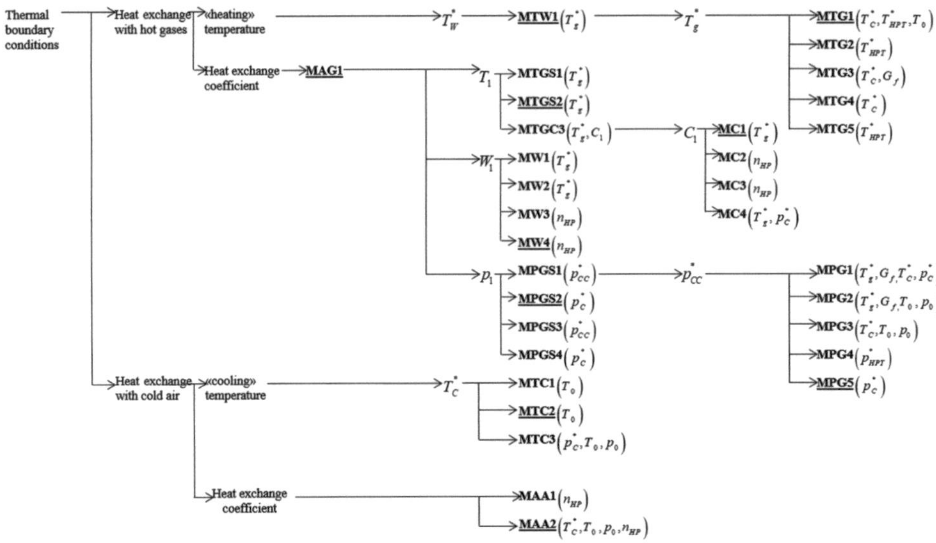

Figure 5.
Structure of the alternative models developed for our test case.

Figure 3, it is necessary to perform an inverse conversion of Eq. (16) to calculate the value of T_g^* as follows:

$$T_g^* = \left[A_{TG1\ cor}(x) \cdot \left(T_{C\ cor}^* - T_0\right) + T_{HPT\ cor}^*\right] \cdot \frac{T_H^*}{T_0} \tag{17}$$

The total error (see Eq. (6)) was the main criteria to assess the accuracy of the developed models. A total of 11 engine conditions (**Table 2**) were considered for the model validation (model robustness analysis).

Figure 6(a) depicts the total error in the prediction of T_g^* using the model MTG1 with different gas path measured parameters as argument in the polynomial to describe the internal model $A_{TG1\ cor}(x)$. From this figure it is clear that the lowest error is obtained when T_{HPT}^* is set as argument.

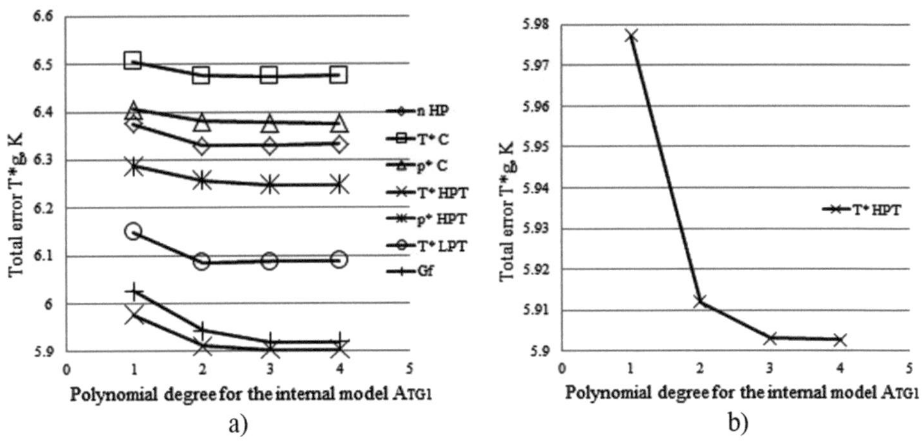

Figure 6.
Total error in the prediction of T_g^* using model MTG1. (a) Using seven different measured parameters as argument; (b) detailed view for the best argument.

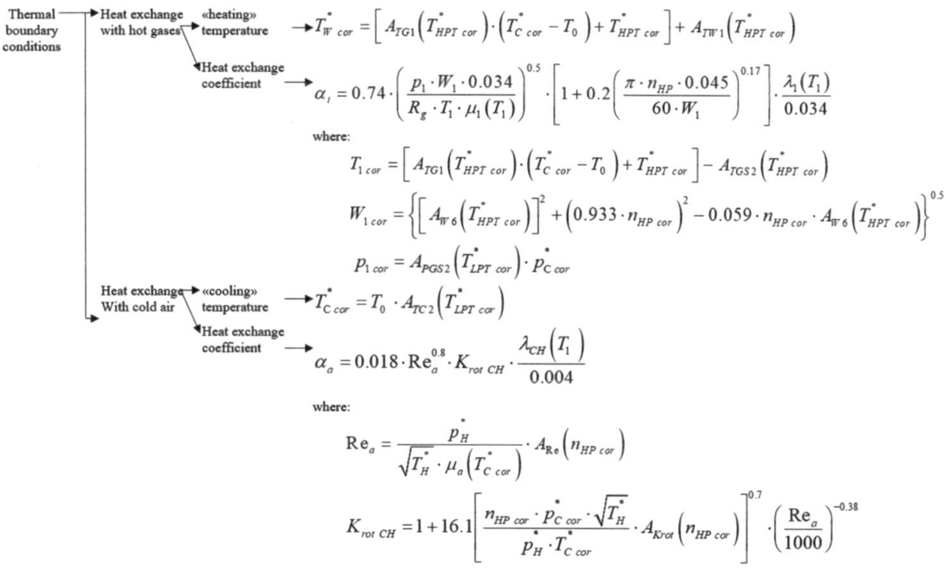

Figure 7.
Selected models to monitor the thermal boundary conditions of the test case.

Figure 6(b) shows that it is sufficient to use a third-degree polynomial as further increment does not reduce the total error in the prediction.

From this analysis the polynomial degree and the gas path measured parameter to be used as argument in Eq. (17) are selected:

$$A_{TG1\ cor}(x) = \left(-6.79 \cdot 10^{-10} \cdot T^{*3}_{HPT\ cor} + 2.19 \cdot 10^{-6} \cdot T^{*2}_{HPT\ cor} - 2.47 \cdot 10^{-} \cdot T^{*}_{HPT\ cor} + 1.87\right) \tag{18}$$

The selection of the argument in the i-internal models and the best polynomial degree for all the developed models was done using the same methodology. After the model verification of all the developed models, the best models were selected. **Figure 7** shows the selected models to calculate the thermal boundary conditions for the test case.

6. Comparative analysis for the model accuracy

Let us conduct a comparative analysis between two approaches for model developing: the first approach for model developing [14] uses the theory of similarity (reference model), and the second approach is proposed in this chapter and uses a physics-based methodology.

6.1 Thermal boundary condition prediction

The thermal boundary conditions for our test case (see **Table 3**) were calculated using models developed with both methodologies.

The total error was calculated according to Eq. (6), and the model robustness analysis took into account the 11 engine health conditions listed in **Table 2**. It is of particular interest to analyze the model robustness, since such analysis for the reference methodology does not exist.

In **Figure 8** the truncation errors for the thermal boundary condition prediction are presented.

From **Figure 8**, it is clear that the models developed using the approach introduced in this chapter are more robust; it means that the models are less sensitive to the deviations from a healthy engine condition. This is a major advantage compared to the reference methodology based in the theory of similarity [14].

6.2 Prediction of the thermal-stress condition and engine lifetime

The thermal-stress engine condition was calculated using Eqs. (1) and (2). The prediction of the turbine blade lifetime is a very complex process, which involves different factors; however, a conservative lifetime prediction will be enough to assess the impact that a new model developing methodology has on the accuracy of the lifetime prediction. According to the author [22], a practical way to predict lifetime is the Larson-Miller relation:

$$t_r = 10^{PLM/t_{cr}-C} \qquad (19)$$

Here *PLM* is the Larson-Muller parameter; *C* is a coefficient, which for the test case is equal to 20.

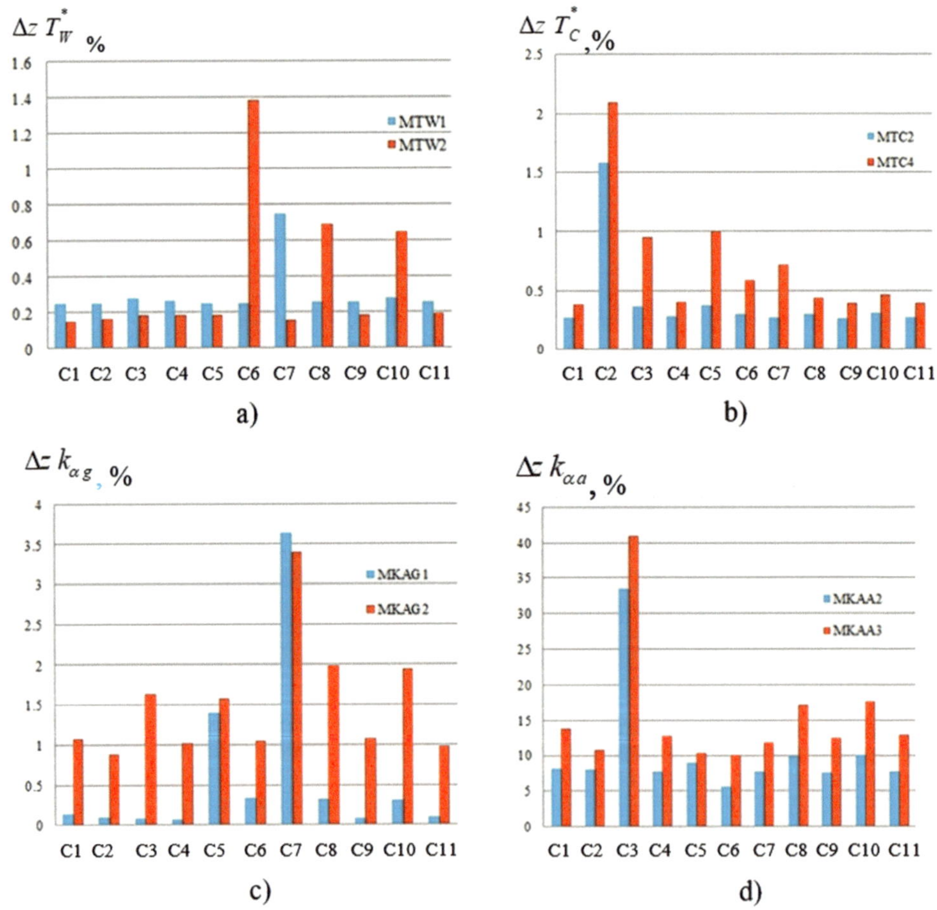

Figure 8.
Truncation error in the prediction of thermal boundary conditions for different engine health conditions. Blue color, physics-based models; red color, models developed on theory of similarity. (a) T_W^*; (b) T_C^*; (c) $k_{\alpha g}$; (d) $k_{\alpha a}$.

As mentioned in Section 5, it is of particular interest to analyze what is the influence of the inclusion or exclusion of the compressor temperature T_C^* as a gas path measured parameter in the accuracy of TC, SC, and lifetime prediction t_r. Therefore, the thermal-stress condition and lifetime prediction are calculated for two cases. For the first case, the compressor temperature is not measured, and for the second case, the compressor temperature is measured.

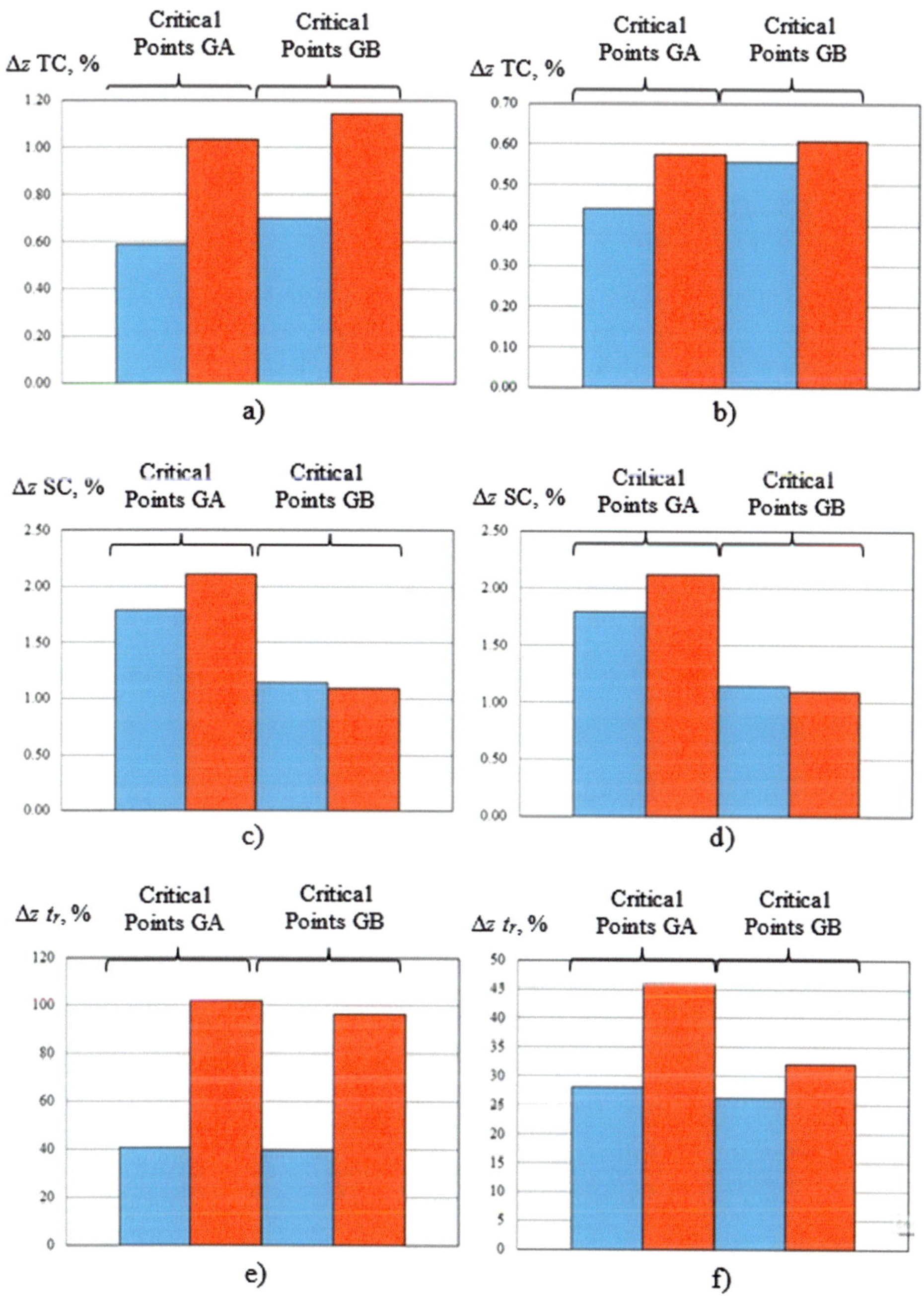

Figure 9.
Total error in the prediction of TC, SC, and t_r. Blue color, physics-based models; red color, models developed on the theory of similarity. (a, c, e) T_C^* is an unmeasured parameter; (b, d, f) T_C^* is a measured parameter.

6.2.1 Prediction of the thermal-stress condition and engine lifetime when the compressor temperature is not measured

Figure 9 presents the total error in the prediction of TC, SC, and lifetime t_r. From **Figure 9(a)** it is clear that the physics-based models give a better prediction of TC. **Figure 9(c)** shows that the improvement in the prediction of SC is not as significant, especially for the critical points GB. **Figure 9(e)** shows that the prediction of the lifetime t_r is highly improved compared to the results obtained when the models based on the theory of similarity are used to calculate the thermal boundary conditions.

From this analysis, it is clear that the methodology used for developing the models of the unmeasured parameters has a great impact in the final accuracy of the lifetime t_r prediction.

6.2.2 Prediction of the thermal-stress condition and engine lifetime when the compressor temperature is measured

We repeated the calculation, but this time the value of the compressor temperature is measured. **Figure 9(b)** shows that the total error in the prediction of TC for both groups is still better when using physics-based models.

Figure 9(d) shows that the accuracy in the prediction of the SC is not significantly affected. From **Figure 9(f)** it is clear that the improvement in the prediction of TC has a significant impact in the prediction of the lifetime t_r, for example, for the critical point GA when using physics-based models, an improvement in the prediction of TC from 0.58 to 0.44% leads to a better lifetime prediction from 45.95 to 27.96%.

We can conclude that the accuracy in the prediction of the lifetime is highly improved when the temperature after the compressor is a measured parameter.

7. Conclusions

A new approach for model developing to estimate the unmeasured parameters in an engine lifetime monitoring system was introduced. This is an effort to increase the accuracy of the lifetime prediction.

All the developed models have a very simple structure and are physics-based, making them ideal to be applied in an on-line lifetime monitoring system.

A turbine blade mounted on the first stage of the high-pressure turbine of a two-spool free turbine power plant is the test case.

Several alternative models were developed using different basic equations. Some of the models include in their structure an internal model, which characterizes the thermodynamic properties of the working fluid, such as efficiencies, pressure loss factors, and others. The internal models were defined by a polynomial function. The best measured parameter used as argument in the polynomial and the degree of the polynomial function were selected using the mean square error as criteria.

All the necessary data for model developing and validation was generated with the engine thermodynamic model.

The truncation and instrumental error are the main criteria to select the best models. Ten engine faulty conditions were considered for the robustness analysis of the models.

A comparative analysis between two model developing methodologies was conducted: physics-based methodology (proposed in this chapter) and models developed on the theory of similarity (reference). The results show that the physics-based models are less sensitive to shifts from the engine healthy condition.

It was found that the use of the proposed model developing methodology provides a better estimation of the thermal boundary conditions, which leads to a significantly better prediction of the lifetime.

It was proven that the compressor temperature has a great impact in the lifetime prediction. If this parameter is not measured, then the accuracy of the lifetime prediction is significantly worse compared to the results obtained when the compressor temperature is measured.

The obtained results show that it is possible to use the proposed model developing methodology in real applications; however, it is necessary to take into account a proper interpretation of the results obtained in this chapter. The reference data, which was used to determine the accuracy of the models, is simulated data; therefore, it is possible that the errors of the lifetime monitoring under real conditions will grow. To avoid such inconvenience, it is possible to replace the data generated with the help of the engine thermodynamic model with real data. This approach is valid, since the engine thermodynamic model is not part of the proposed model developing methodology. In order to obtain the necessary real data, it will be necessary to use additional instrumentation under engine test bed conditions.

Nomenclature

Subscripts

a	air
cor	corrected parameter
cr	critical point
C	compressor
CC	combustion chamber
CH	channel
f	fuel
g	gas
H	atmospheric conditions
HP	high pressure
HPT	high-pressure turbine
i	i-engine mode
INS	instrumental error
j	j-engine health condition
LPT	low-pressure turbine
m	value calculated with the help of models, mechanical
NB	nozzle box
ref	reference engine mode
ST	gas pumping station
$S1$	cooling
$S2$	heating
t	thermal
T	turbine
TR	truncation error
TT	total mean square error
W	relative velocity

Designations

G	consumption
T	temperature
p	pressure
n	rotational speed, number of engine health conditions
t	blade temperature
k	coefficient, isentropic factor
$\overline{S}$	dimensionless parameter
z	unmeasured parameter
Y	gas path measured parameters
W	unmeasured parameters describing thermodynamic properties of the working fluid
A	internal model
x	measured parameter, argument of polynomial
p_0	standard atmospheric pressure (101.3 KPa)
T_0	standard atmospheric temperature (288.15 K)
y	measured parameter
N	sample size (number of engine operation modes), power
C	engine condition, coefficient
L	thermodynamic work
C_p	specific heat at constant pressure
A	coefficient
Re	Reynolds number
K_α	relation of heat transfer coefficient at current and at a reference operation mode
t_r	lifetime

Greek symbols

*	stagnation parameter
α	heat transfer coefficient
Θ	dimensionless parameter
σ	stress, mean square error, total pressure conservation
η	efficiency
δ	shift from healthy engine condition
μ	dynamic viscosity

Author details

Cristhian Maravilla* and Sergiy Yepifanov
National Aerospace University of Ukraine, Kharkiv, Ukraine

*Address all correspondence to: cris_mahe@hotmail.com

References

[1] Jin H, Lodwen P, Pistor R. Remaining life assessment of power turbine disks. In: Proceedings of ASME Turbo Expo 2008: Power for Land, Sea and Air. Berlin, Germany: GT2008-51010; 2008

[2] Alexander V. Gas turbine plant thermal performance degradation assessment. In: Proceedings of ASME Turbo Expo 2008: Power for Land, Sea and Air. Berlin, Germany: GT2008-50032; 2008

[3] Phillip D, David D. Remaining life assessment technology applied to steam turbines and hot gas expanders. In: Proceedings of ASME Turbo Expo 2011: Power for Land, Sea and Air. Vancouver, Canada: GT2011-45324; 2011

[4] Parthasarathy G, Menon S, Richardson K. Neural network models for usage based remaining life computation. In: Proceedings of ASME Turbo Expo 2006: Power for Land, Sea and Air. Barcelona, Spain: GT2006-91099; 2006

[5] Yoon J, Kim T, Lee J. Simulation of performance deterioration of a microturbine and application of neural network to its performance diagnosis. In: Proceedings of ASME Turbo Expo 2008: Power for Land, Sea and Air. Berlin, Germany: GT2008-51494; 2008

[6] Fast M, Assadi M, De S. Condition based maintenance of gas turbines using simulation data and artificial neural networks: A demonstration of feasibility. In: Proceedings of ASME Turbo Expo 2008: Power for Land, Sea and Air. Berlin, Germany: GT2008-50768; 2008

[7] Palme T, Breuhaus P, Assadi M. Early warning of gas turbine failure by nonlinear feature extraction using an auto-associative neural network approach. In: Proceedings of ASME Turbo Expo 2011: Power for Land, Sea and Air. Vancouver, Canada: GT2011-45991; 2011

[8] Blumenthal R, Hutchinson B, Zori L. Investigation of transient CFD methods applied to a transonic compressor stage. In: Proceedings of ASME Turbo Expo 2011: Power for Land, Sea and Air. Vancouver, Canada: GT2011-46635; 2011

[9] Weiss J, Subramanian V, Hall K. Simulation of unsteady turbomachinery flows using an implicitly coupled nonlinear harmonic balance method. In: Proceedings of ASME Turbo Expo 2011: Power for Land, Sea and Air. Vancouver, Canada: GT2011-46367; 2011

[10] Jeromin A, Eichler C, Noll B. Full 3D conjugate heat transfer simulation and heat transfer coefficient prediction for the effusion-cooled warm of a gas turbine combustor. In: Proceedings of ASME Turbo Expo 2008: Power for Land, Sea and Air. Berlin, Germany: GT2008-50422; 2008

[11] Giuseppe F, Carlevaro F. Gas turbine maintenance policy: A statistical methodology to prove interdependency between number of starts and running hours. In: Proceedings of ASME Turbo Expo 2002: Power for Land, Sea and Air. Amsterdam, Netherlands: GT2002-30281; 2002

[12] Eshati S, Abdul Ghafir M, Laskaridis P, Li Y. Impact of operating conditions and design parameters on gas turbine hot section creep life. In: Proceedings of ASME Turbo Expo 2010: Power for Land, Sea and Air. Glasgow, UK: GT2010-22334; 2010

[13] Maccio M, Rebizzo A, Traversone L, Bordo L. Rotor components life evaluation validated by field operation data. In: Proceedings of ASME Turbo

Expo 2010: Power for Land, Sea and Air. Glasgow, UK: GT2010-22741; 2010

[14] Oleynik A. Concepts and development of methods for lifetime monitoring based on the identification of the dynamics of the thermal-stress conditions of aviation GTE critical components [thesis]. National Aerospace University of Ukraine; 2006

[15] Jiang L, Eli H, Hee-Koo M. 3D rams prediction of gas-side heat transfer coefficients on turbine blade and end wall. In: Proceedings of ASME Turbo Expo 2011: Power for Land, Sea and Air. Vancouver, Canada: GT2011-46723; 2011

[16] Abubakar M, Reyad A, Gordon E, John E. Conjugate heat transfer CFD predictions of impingement heat transfer: Influence of the number of holes for a constant pitch to diameter ratio X/D. In: Proceedings of ASME Turbo Expo 2014: Power for Land, Sea and Air. Dusseldorf, Germany: GT2014-25268; 2014

[17] Wieslaw B, Zhong Z, David D, Chen W, Wu X. Residual life assessment of a critical component of a gas turbine – Achievements and challenges. In: Proceedings of ASME Turbo Expo 2014: Power for Land, Sea and Air. Dusseldorf, Germany: GT2014-26423; 2014

[18] Yepifanov S, Kuzmenko B, Ryumshin N. Synthesis of Turbine Engine Control and Diagnosing Systems. Kiev, Ukraine: Technical Publishing; 1998

[19] Maravilla C, Yepifanov S, Loboda I. Improved turbine blade lifetime prediction. In: Proceedings of ASME Turbo Expo 2015: Power for Land, Sea and Air. Montreal, Canada: GT2015-43046; 2015

[20] Kulagin B. Theory and Design of Aircraft Engines and Power Pants. Moscow: Mashinostroenie; 2003. p. 616

[21] Maravilla C. Improvement in the prediction of the thermal boundary conditions in a turbine blade monitoring system [thesis]. National Aerospace University of Ukraine; 2016

[22] Krykonov D. Handbook of Engine Lifetime Prediction. Kharkiv, Ukraine: National Aerospace University of Ukraine; 2005. p. 67

Section 3

Wind Turbines

Chapter 5

Modeling and Simulation of the Variable Speed Wind Turbine Based on a Doubly Fed Induction Generator

Imane Idrissi, Houcine Chafouk, Rachid El Bachtiri and Maha Khanfara

Abstract

This chapter presents the modeling and simulation results of variable speed wind turbine driven by doubly fed induction generator (DFIG). The feeding of the generator is ensured through its stator directly connected to the electrical grid and by its rotor connected to the grid through two power converters, which are controlled by the pulse width modulation (PWM) technique. This configuration is the most used in the wind power generation systems. For the variable speed operation of the studied system, the maximum power point tracking strategy is applied for the turbine, and the stator flux-oriented vector control is used for the generator. The MATLAB/Simulink software is used for the system modeling and simulation. For the wind velocity model, a random wind profile is simulated, and the turbine and the generator parameters are extracted from an existing wind turbine system in the literature. The obtained results are addressed in this chapter.

Keywords: wind turbine, variable speed operation, DFIG, PWM, MATLAB/Simulink, MPPT

1. Introduction

With the global warming issues and the climate changes, there is a serious need for the use of the renewable energy resources in the electricity generation industry. Currently, the wind represents one of the most important renewable energy resources, used for generating electrical energy in the world. In terms of the total installed wind capacity, it becomes up to 539 GW across the globe in 2017 [1]. The rapid rate of the wind energy industry growth is caused by the cost-effectiveness of electricity production from wind farms, compared to electricity production cost from fossil fuel energy [2], the stability of electricity cost [3], the short commissioning time of wind farms [4], and the ingenuity of skillful engineers.

According to a wind market survey, the doubly fed induction generator (DFIG) is the most popular generator used in the speed variable wind turbines (SVWT) [5]. It is a wound rotor asynchronous machine which has the stator windings directly connected to the electrical grid, and its rotor is linked to the constant frequency grid by means of two bidirectional power converters.

This configuration, known as "Scherbius structure" and shown in **Figure 1**, has several advantages of controlling independently active and reactive power [6, 7]. Moreover, the power converters used are sized to transfer only a fraction, equal to at most 30% of the turbine rated power [8, 9], which results in small-size, low-cost, less acoustic noise and reduced loss rate in the power converters [10]. Moreover, The DFIG-based wind turbine allows the rotor speed to be varied with the wind speed, and the speed variation range is around ±30% around the synchronism speed [11]. As a result, the wind generation system could operate in hyposynchronous and hypersynchronous mode, which would extract the maximum aerodynamic power for each wind speed value.

In order to design fault diagnosis and control approaches based on models for wind turbines, the development of a mathematical model, which represents as much details as technically possible and gives an accurate idea of the dynamic behavior of the system, seems to be an important step. For several purposes, different wind turbine models have been developed. In literature, we find for wind systems the aerodynamic model, which aims to verify and optimize the blade design, depending on predefined specifications, while the mechanical model is used by engineers for establishing a safe and economical dimensioning of the whole wind turbine system. Moreover, the economic model is used in the case of manufacturing and installing wind turbines with the purpose to evaluate the cost-effectiveness. In addition, there are models which predict the weather conditions and the power output of wind farms. Furthermore, there are models, which have the objective of evaluating the impact of wind turbines on the environment such as the evaluation of noise produced by the wind turbine operation. Finally, the general-purpose models concerned with the electrical properties of wind turbines are widely used.

The speed variable wind turbine (SVWT) model, developed and simulated in this work, is concerned with providing time simulation signals that can be exploited for designing fault diagnosis approaches based on models; the software tool used for simulation is the MATLAB/Simulink environment. In [12], a wind turbine model of a fixed-speed, stall-regulated system has been developed with the aim of measuring and evaluating the power quality impact of wind turbines on the grid. In addition, a model for a wind turbine generation system based on a DFIG, including the mechanical dynamics, the wind turbine electrical system, the converter, and the electrical grid has been presented in [13]. Luis et al. [14] presented the most commonly used wind turbine model meeting objectives as production energy, safety of turbine, grid connection, and others. Moreover, in [15], the detailed

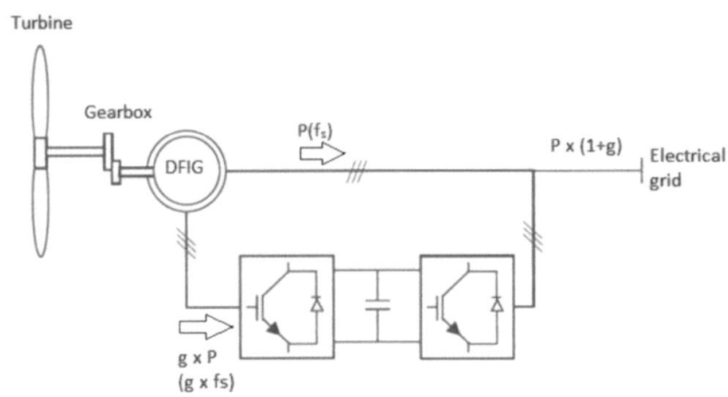

Figure 1.
Architecture Scherbius of DFIG-based wind turbine.

mechanical structural modeling of the wind turbine connected to the grid and based on DFIG has been developed, and it has been validated using NREL's simulation tool, FAST v7; for analysis of the dynamic behavior of the wind power plant with DFIG under the grid fault conditions, modeling of the whole system has been established in [16]. Furthermore, with the use of an electromagnetic transient simulation software, the wind turbine driven by DFIG model is elaborated in [17].

The organization of this chapter is as follows: the wind turbine structure is described in Section 2; the wind modeling turbine is presented in Section 3. In Section 4, the simulation results in MATLAB/Simulink environment are shown. Finally, the conclusion is presented in Section 5.

2. The wind turbine structure

The wind turbine is a complex system containing different components, which involved different domains: electrical, mechanical, and electronic areas and others. The complete model of the studied wind turbine system is represented by a set of blocks each representing a functional entity of the system. The general structure of a wind system is given by **Figure 2**.

As it is shown in the **Figure 2**, the wind velocity and the fixed-pitch angle represent the input of the wind turbine system. The aerodynamic conversion entity is formed of three blades which capture and convert the kinetic energy of the wind into mechanical energy, recovered on the slow rotating shaft. Then, the gearbox device increases the turbine low speed and makes it suitable with the generator rotational speed, which is about 1500 rpm. The generator receives mechanical energy and transforms it into electrical energy. The two power converters used are insulated gate bipolar transistor (IGBT) type and controlled by pulse width modulation (PWM) technique; they allow the independent control of the active and reactive powers and also the transfer of the slip power in two directions: from the generator to the network and from the network to the generator.

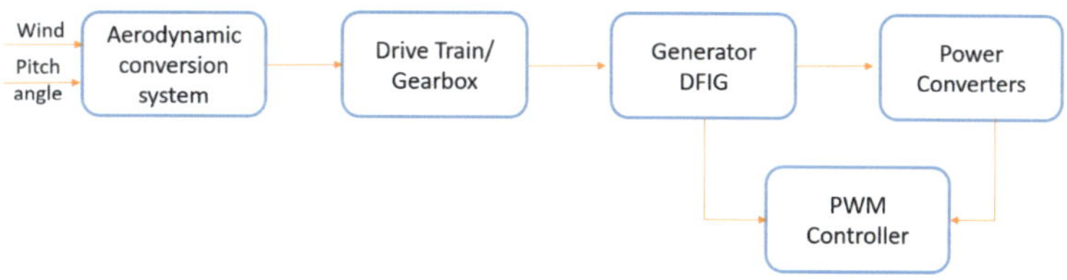

Figure 2.
Wind turbine system structure.

3. The wind turbine modeling

In this section, the mathematical model of each wind turbine block is presented.

3.1 Wind speed model

The wind resource is an important element in a wind energy system, and it represents a determining factor in the calculation of electricity production because, under optimal conditions, the power captured by the wind turbine is a cubic function of the wind speed. The wind is a moving air mass, and the wind kinetic energy is given by:

$$E = \frac{1}{2} \cdot m \cdot v^2 \tag{1}$$

where m is the moving air mass [g] and v is the air moving speed [m/s]. The wind power during is expressed as:

$$P_{wind} = \frac{E}{\Delta t} \tag{2}$$

The wind speed v is generally represented by a scalar function evolving over time, given by V = f(t). It can also be divided into two components: a slowly varying part denoted as V_0 and a random varying part denoted as V_t; it represents the wind fluctuations. Therefore, the wind velocity can be written as:

$$V(t) = V_0 + V_t(t) \tag{3}$$

To mathematically model the wind speed profile, the literature offers three techniques:

- The first method is white noise filtering technique, in which the turbulence impact is corrected by the use of a low-pass filter having the following transfer function [18]:

$$F(s) = \frac{1}{1 + \tau \cdot s} \tag{4}$$

where τ is the filter time constant. It depends on the rotor diameter and the wind turbulence intensity and the average wind speed. **Figure 3** shows the method of reconstruction of the wind profile using this technique.

- The second method of generating the wind speed profile is that which describes wind variations using the spectral density established by meteorologist I. Van der Hoven. In this model, the turbulence part is considered as a stationary random process, and therefore it does not depend on

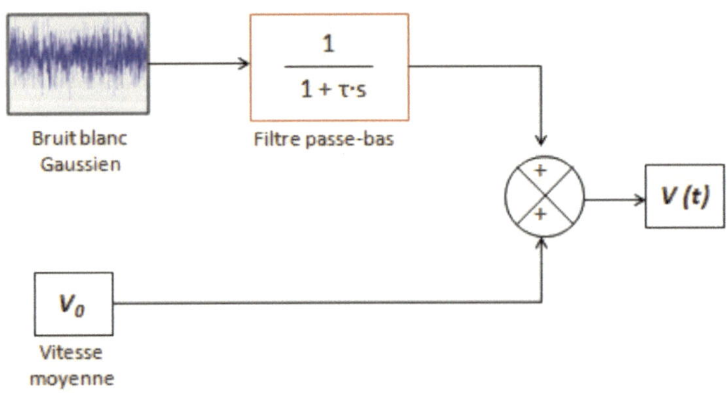

Figure 3.
Wind profile construction scheme by white noise filtering.

the variation of the mean wind speed [19, 20]. The variation of the wind speed v(t) is thus written in the form of the harmonic sum:

$$V_v(t) = A + \sum_{i=1}^{n} a_k \cdot \sin(\omega_k \cdot t) \tag{5}$$

where A is the wind speed average value; a_k is the amplitude of k-order harmonic; ω_k is the pulsation of k-order harmonic; and i is the last harmonic rank retained in the wind profile calculation.

- The third method is the Weibull distribution in which a given site wind potential is obtained by measuring the average wind speed in regular time intervals. Then, the data obtained are then divided into numbers by wind speed classes using histogram [21]. The wind profile over a desired time period, respecting the Weibull distribution, is given by:

$$V_v(t) = (1 + \xi_v(t) - \xi_v) \cdot V_v \tag{6}$$

where V_v is the wind speed average value and ξ_v is the disturbance mean value expressed by:

$$\xi_v(t) = \left(-\frac{ln(rand(t))}{c_v}\right)^{\frac{1}{k_v}} \tag{7}$$

where rand(t) is a function generating, in a uniform distribution, random numbers between 0 and 1 and (C_v, k_v) is a parameter pair, determined by analysis of the wind class histogram. C_v is a scale factor generally greater than 5. The shape factor k_v is greater than 3 if the histogram shape is like that of a normal distribution, characterized by an uniform distribution around a mean value.

In this work, we adopted the second method to generate the random profile of the wind speed applied in the studied wind system input.

3.2 Aerodynamic conversion model

The aerodynamic conversion system is the wind turbine part, which is facing the wind; it generally comprises three blades of length R. Three-bladed wind turbines are much more common than two-bladed wind turbines. The turbine captures the kinetic energy of the wind and transforms it into mechanical energy recovered on the slow rotating shaft.

The kinetic power of the wind is given by.

$$P_{wind} = \frac{\rho \cdot S \cdot v_{wind}^3}{2} \tag{8}$$

The aerodynamic power is expressed as follows:

$$P_{aero} = C_p(\lambda, \beta) \cdot P_{wind} = C_p(\lambda, \beta) \cdot \frac{\rho \cdot S \cdot v_{wind}^3}{2} \tag{9}$$

The aerodynamic torque T_{aer} is given by the following expression:

$$T_{aer} = \frac{P_{aero}}{\Omega_t} = \frac{1}{2 \cdot \Omega_t} \cdot C_p(\lambda, \beta) \cdot \rho \cdot S \cdot v_{wind}^3 \tag{10}$$

where Ωt is the turbine speed [rad/s], ρ is the air density, ρ = 1.225 kg/m^3, S = π R_2 is the rotor surface [m^2], R is the blade length [m], and v_{wind} is the wind speed upstream of the wind turbine rotor [m/s]. λ is the speed ratio. It is a unitless parameter, related to the design of each wind turbine, and it represents the ratio between the speed of the blade's end and that of the wind at the rotor axis or also called hub. λ is expressed as follows:

$$\lambda = \frac{\Omega_t \cdot R}{v} \tag{11}$$

This parameter depends on the blade number of the wind turbine. If the blade number is reduced, the rotor speed is high, and a maximum of power is extracted from the wind. In the case of multiblade wind turbines (Western Wind Turbines), the speed ratio is equal to 1; for wind turbines with a single blade, λ is about 11. The three-bladed wind turbines, as in our study, have a speed ratio of 6 to 7. The speed ratio of Savonius wind turbines is less than 1 [22].

Cp is the power coefficient or aerodynamic transfer efficiency that varies with the wind speed. This coefficient has no unit, and it depends mainly on the blade aerodynamics, the speed ratio λ, and the blade orientation angle β. Betz has determined a theoretical maximum limit of the power coefficient C_{pmax} = 16/27 ∽ 0.59. Taking into account losses, wind turbines never operate at this maximum limit, and the best-performing wind turbines have a Cp between 0.35 and 0.45. Cp is specific to each wind turbine, and its expression is given by the wind turbine manufacturer or using nonlinear formulas. To calculate the coefficient Cp, different numerical approximations have been proposed in the literature. The Cp expressions frequently encountered in the literature are presented in **Table 1**.

Since we had as an objective the modeling and simulation of a three-bladed wind turbine with a nominal power of 3 kW; the parameters of both: the wind turbine and the generator have been used from [30]. For this reason, the analytical expression of the power coefficient C_p is given by:

$$C_p = 6 \cdot 10^{-7} \cdot \lambda^5 + 10^{-5} \cdot \lambda^4 - 65 \cdot 10^{-5} \cdot \lambda^3 + 2 \cdot 10^{-5} \cdot \lambda^2 + 76 \cdot 10^{-3} \cdot \lambda + 0.007 \tag{12}$$

This coefficient has a maximum value equal to 0,35 ($C_{pmax} = 0{,}35$) and an optimal value of relative speed equal to 7 (λ = 7).

The block diagram presenting the aerodynamic part is shown in **Figure 4**.

3.3 Gearbox model

The mechanical part of the wind turbine consists of the turbine shaft rotating slowly at speed Ω_t, the gearbox having the multiplication gain G and driving the generator at a speed Ω_g, by means of a fast secondary shaft.

The gearbox is a device that allows to multiply the turbine speed of Ω_t by a multiplication gain G to make it adapt to the rapid speed of the generator Ω_g. This device is considered ideal, because the gearbox elasticity, friction, and energy losses are considered negligible. The two equations mathematically modeling the operation of this device are given as follows:

$$\begin{cases} T_g = \dfrac{T_{aer}}{G} \\ \Omega_t = \dfrac{\Omega_g}{G} \end{cases} \tag{13}$$

DOI: http://dx.doi.org/10.5772/intechopen.83690

Power coefficient type, Cp	Formula
Exponential	$0.22 \cdot \left[\frac{116}{\lambda_i} - 0.4 \cdot \beta - 5\right] \cdot e^{\frac{12.5}{\lambda_i}}$ avec $\frac{1}{\lambda_i} = \frac{1}{\lambda + 0.08 \cdot \beta} - \frac{0.035}{\beta^3 + 1}$ [23]
	$0.5 \cdot \left[\frac{116}{\lambda_i} - 0,4 \cdot \beta - 5\right] \cdot e^{-\frac{21}{\lambda_i}} + 0.068 \cdot \lambda$ [8, 24, 25] Avec: $\frac{1}{\lambda_i} = \frac{1}{\lambda + 0.08 \cdot \beta} - \frac{0.035}{\beta^3 + 1}$
	$0.5176 \cdot \left[\frac{115}{\lambda_i} - 0.4 \cdot \beta - 5\right] \cdot e^{-\frac{21}{\lambda_i}} + 0.0068 \cdot \lambda$ [26] Avec: $\frac{1}{\lambda_i} = \frac{1}{\lambda + 0.08 \cdot \beta} - \frac{0.035}{\beta^3 - 1}$
	$0.5109 \cdot \left[\frac{116}{\lambda_i} - 0.4 \cdot \beta - 5\right] \cdot e^{-\frac{21}{\lambda_i}} + 0.0068 \cdot \lambda$ [27] Avec: $\frac{1}{\lambda_i} = \frac{1}{\lambda + 0.08 \cdot \beta} - \frac{0.035}{\beta^3 - 1}$
	$0.44 \cdot \left[\frac{125}{\lambda_i} - 6.94\right] \cdot e^{-\frac{16.5}{\lambda_i}}$ [21] Avec: $\lambda_i = \frac{1}{\lambda + 0,002}$
Sinusoidal	$0.5 - 0.167 \cdot (\beta - 2) \cdot \sin\left[\frac{\pi \cdot (\lambda - 3)}{18.9 - 0.3 \cdot (\beta - 2)}\right] - 0.00184 \cdot (\lambda - 3) \cdot (\beta - 2)$ [28]
	$0.3 - 0.00167 \cdot (\beta - 2) \cdot sin\left[\frac{\pi \cdot (\lambda + 0,1)}{10 - 0.3 \cdot (\beta - 2)}\right] - 0.00184 \cdot (\lambda - 3) \cdot \beta$ [29]
Polynomial	$6 \cdot 10^{-7} \cdot \lambda^5 + 10^{-5} \cdot \lambda^4 - 65 \cdot 10^{-5} \cdot \lambda^3 + 2 \cdot 10^{-5} \cdot \lambda^2 + 76 \cdot 10^{-3} \cdot \lambda + 0.007$ [30]
	$79.5633.10^{-5} \cdot \lambda^{-5} - 17.375 \cdot 10^{-4} \cdot \lambda^4 + 9.86 \cdot 10^{-3} \cdot \lambda^3 - 9.4 \cdot 10^{-3} \cdot \lambda^2 + 6.38 \cdot 10^{-2} \cdot \lambda + 0.001$ [31]

Table 1.
Different numerical formulas of the power coefficient Cp.

where Tg is torque on the generator shaft (N·m), T_{aer} is the aerodynamic torque of the wind turbine ($N \cdot m$), Ω_g is the speed generator shaft ($rad \cdot s^{-1}$), Ω_t is the turbine speed shaft ($rad \cdot s^{-1}$), and G is the multiplication gain; it is given by G = N1/N2.

Figure 5 shows the gearbox model for determining the multiplication gain G, and **Figure 6** shows the gearbox block diagram.

The total inertia J consists of the turbine inertia J_t and the generator inertia J_g; it can be written according to the following equation [28]:

$$J = \frac{J_t}{G^2} + J_g \tag{14}$$

The total viscous friction coefficient f_v consists of the generator friction coefficient f_g and the turbine friction coefficient f_t. The coefficient f_v can be expressed as follows:

$$f_v = \frac{f_t}{G^2} + f_g \tag{15}$$

Therefore, the mechanical part can be modeled according to the diagram shown in **Figure 7**.

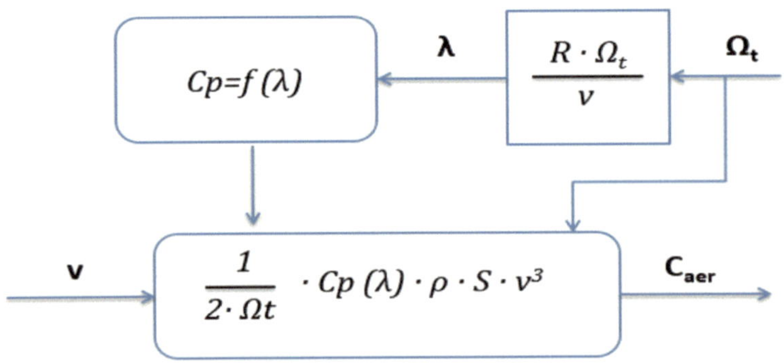

Figure 4.
Block diagram of the aerodynamic part.

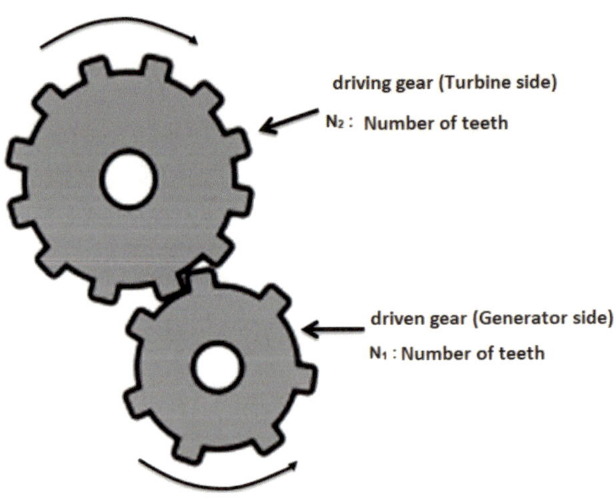

Figure 5.
The gearbox model.

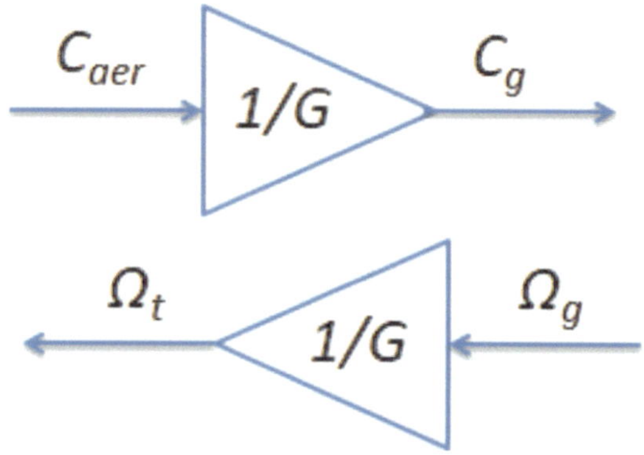

Figure 6.
Block diagram of the gearbox.

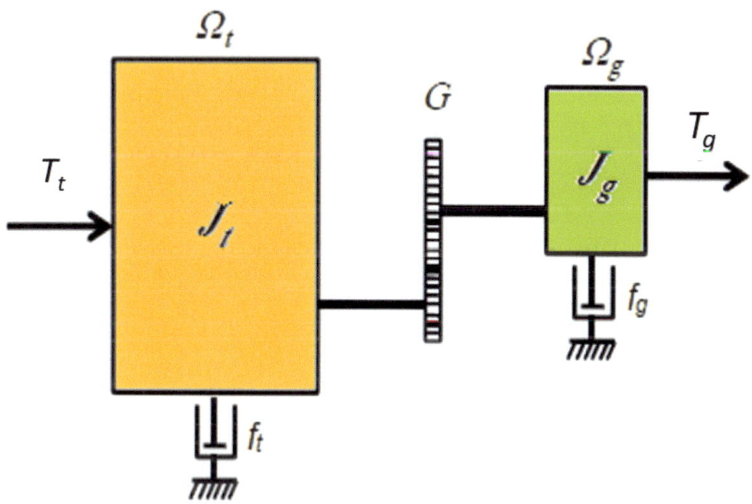

Figure 7.
The several coefficients of the wind turbine mechanical part.

The generator speed Ω_g depends on the total mechanical torque T_{mec}. This torque is the result of the electromagnetic torque of the generator T_{em}, the viscous friction torque T_v, and the torque applied on the generator shaft T_g.

$$T_{mec} = J \cdot \frac{d\Omega_g}{dt} \tag{16}$$

$$T_{mec} = T_g - T_{em} - T_v \tag{17}$$

$$T_v = f \cdot \Omega_g \tag{18}$$

Therefore, from these previously established equations, the differential equation of the mechanical system dynamics is expressed by:

$$J \cdot \frac{d\Omega_g}{dt} = T_g - T_{em} - T_{vis} \tag{19}$$

The block diagram of the wind turbine mechanical part is presented in **Figure 8**. The diagram block of the whole wind turbine system is given in **Figure 9**.

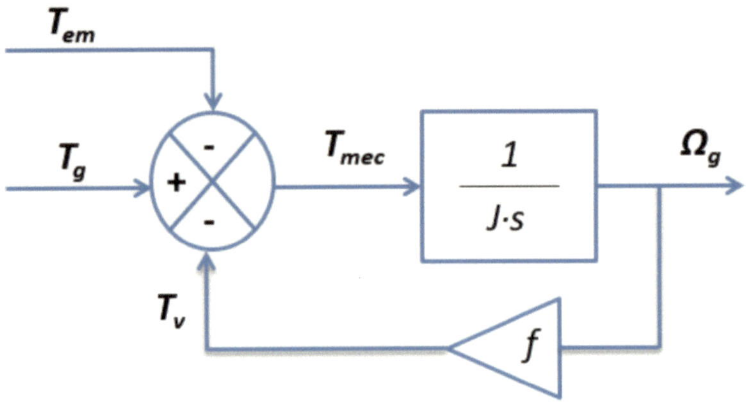

Figure 8.
Block diagram of the wind turbine mechanical part.

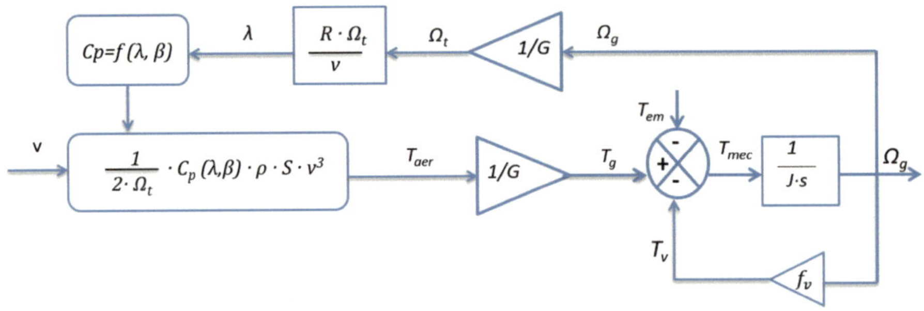

Figure 9.
Block diagram of the whole wind turbine system.

In order to continuously reach the maximum power point provided by a wind turbine, operating over a wide range of wind speed, the maximum power point tracking (MPPT) control technique is used. In this chapter, the MPPT control without controlling the mechanical speed is presented [32]. This control strategy is based on the assumption that the wind speed little varies in steady state compared to the electrical constants of the wind turbine system. Therefore, at the maximum power point, the relative speed λ is equal to its optimum value λ_{opt}, and the power coefficient C_p is equal to its maximum value $C_{p\text{-}max}$, while the reference electromagnetic torque C^*_{em} is given by:

$$C^*_{em} = \frac{C_{aer-est}}{G} \tag{20}$$

$$C^*_{em} = C_{p-max} \cdot \frac{\rho \cdot \pi \cdot R^5 \cdot \Omega_g^2}{2 \cdot \lambda_{opt}^3 \cdot G^3} \tag{21}$$

For simplification, the parameter K is expressed as:

$$K = C_{p-max} \cdot \frac{\rho \cdot \pi \cdot R^5}{2 \cdot \lambda_{opt}^3 \cdot G^3} \tag{22}$$

Therefore:

$$C^*_{em} = K \cdot \Omega_g^2 \tag{23}$$

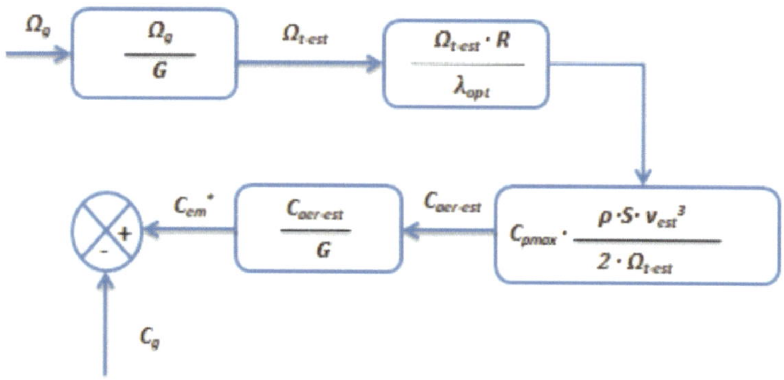

Figure 10.
Block diagram of the MPPT control without mechanical speed.

The reference electromagnetic torque is proportional to the square of the generator speed Ω_g. The block diagram which presents the MPPT control strategy without the measurement of the generator speed is shown in **Figure 10**.

3.4 DFIG model

The doubly fed induction generator (DFIG) is a three-phase asynchronous machine, powered by two sources: by its stator and its rotor at the same time. Its main advantage is that it offers the possibility of controlling the power flows for the hypo- and hypersynchronous modes, either in the motor or generator operation. It also allows the variable speed operation of the system where it is integrated.

The DFIG model in the stationary reference frame, noted (α, β), is given in the state representation [33] as follows:

$$\frac{d}{dt}\begin{bmatrix} i_{s\alpha} \\ i_{s\beta} \\ i_{r\alpha} \\ i_{r\beta} \end{bmatrix} = [A]\cdot\begin{bmatrix} i_{s\alpha} \\ i_{s\beta} \\ i_{r\alpha} \\ i_{r\beta} \end{bmatrix} + [B]\cdot\begin{bmatrix} v_{s\alpha} \\ v_{s\beta} \\ v_{r\alpha} \\ v_{r\beta} \end{bmatrix} \tag{24}$$

where $i_{s\alpha}$ and $i_{s\beta}$ are the stator currents in the stationary reference frame (α, β); $i_{r\alpha}$ and $i_{r\beta}$ are the rotor currents in the reference frame (α, β); $v_{s\alpha}$ and $v_{s\beta}$ are the stator stresses in the stationary reference frame (α, β); $v_{r\alpha}$ and $v_{r\beta}$ are the rotor voltages in the stationary reference frame (α, β).

The matrices $A \in \mathbb{R}^{n\times n}$, $B \in \mathbb{R}^{n\times m}$, and $C \in \mathbb{R}^{p\times n}$ are, respectively, the state matrix, the input or control matrix, and the output or observation matrix. They are, respectively, given by:

$$A = \begin{bmatrix} \frac{-R_s}{\sigma\cdot L_s} & \left(\frac{(1-\sigma)}{\sigma}\cdot\omega+\omega_s\right) & \frac{M\cdot R_r}{\sigma\cdot L_s\cdot L_r} & \frac{M}{\sigma\cdot L_s}\omega \\ -\left(\frac{(1-\sigma)}{\sigma}\cdot\omega+\omega_s\right) & \frac{-R_s}{\sigma\cdot L_s} & -\frac{M}{\sigma\cdot L_s}\cdot\omega & \frac{M\cdot R_r}{\sigma\cdot L_s\cdot L_r} \\ \frac{M\cdot R_s}{\sigma\cdot L_s\cdot L_r} & -\frac{M}{\sigma\cdot L_r}\cdot\omega & \frac{-R_r}{\sigma\cdot L_r} & \left(\omega_s-\frac{\omega}{\sigma}\right) \\ \frac{M}{\sigma\cdot L_r}\cdot\omega & \frac{M\cdot R_s}{\sigma\cdot L_s\cdot L_r} & -\left(\omega_s-\frac{\omega}{\sigma}\right) & \frac{-R_r}{\sigma\cdot L_r} \end{bmatrix}$$

$$B = \begin{bmatrix} \frac{1}{\sigma \cdot L_s} & 0 & -\frac{M}{\sigma \cdot L_s \cdot L_r} & 0 \\ 0 & \frac{1}{\sigma \cdot L_s} & 0 & -\frac{M}{\sigma \cdot L_s \cdot L_r} \\ -\frac{M}{\sigma \cdot L_s \cdot L_r} & 0 & \frac{1}{\sigma \cdot L_r} & 0 \\ 0 & -\frac{M}{\sigma \cdot L_s \cdot L_r} & 0 & \frac{1}{\sigma \cdot L_r} \end{bmatrix}$$

$$C = \begin{bmatrix} 1 & 0 & 0 & 0 \\ 0 & 1 & 0 & 0 \\ 0 & 0 & 1 & 0 \\ 0 & 0 & 0 & 1 \end{bmatrix}$$

where R_s and L_s are, respectively, the single-phase resistance and the cyclic single-phase inductance of the stator winding; R_r and L_r are, respectively, the single-phase resistance and the cyclic single-phase inductance of the rotor winding; M is the mutual inductance between the stator phase and the rotor phase; $\sigma = 1 - \frac{M2}{L_s.L_r}$ is the leakage coefficient or the Blondel coefficient; ω_s is the synchronism angular speed [rad/s]; and ω is the mechanical angular speed [rad/s].

In order to generate the reference rotor voltages which will be the input of the machine side converter, the stator flux-oriented vector control is applied to DFIG system [34].

3.5 Power converter models

The power electronic converters used consist of a rectifier made using semi-conductors controlled at the opening and closing, and a three-phase voltage inverter consists of three reversible current switch arms, controlled at the opening and closing in the same time. Each arm consists of two switches, which contain each one insulated gate bipolar transistor (IGBT) and an antiparallel diode. The voltage capacitor DC allows the storage of the output rectifier energy. The passive filter type (L, R) is used to connect the inverter to the grid. Both converters used are controlled using pulse width modulation (PWM), and the power converter structure is given in **Figure 11**.

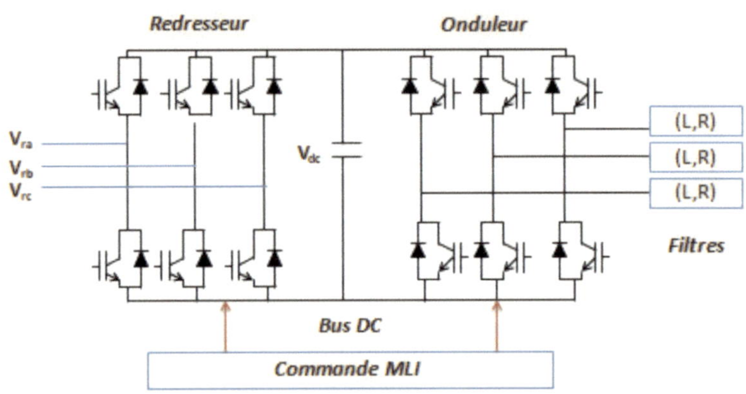

Figure 11.
Structure of the power converters (IGBT).

The input voltages of single phases of the rotor side converter (RSC) are described as follows:

$$\begin{cases} V_{Sa} = \dfrac{2 \cdot S_a - S_b - S_c}{3} \cdot V_{dc} \\ V_{Sb} = \dfrac{2 \cdot S_b - S_a - S_c}{3} \cdot V_{dc} \\ V_{Sc} = \dfrac{2 \cdot S_c - S_a - S_b}{3} \cdot V_{dc} \end{cases} \tag{25}$$

where S_i represents the switch states, supposedly ideal to facilitate the rectifier modeling, defined by:

$$S_i = \begin{cases} 1, \overline{S_i} = 0 \\ 0, \overline{S_i} = 1 \end{cases} ; \mathrm{i} = \mathrm{a, b, c} \tag{26}$$

The rotor voltage equations and the DC capacitor equation are given, respectively:

$$[V_{rabc}] = R_r \cdot [i_{rabc}] + L_r \cdot \frac{d}{dt}[i_{rabc}] + [V_{Sabc}] \tag{27}$$

$$C \cdot \frac{dV_{dc}}{dt} = (S_a \cdot i_{Sa} + S_b \cdot i_{Sb} + S_c \cdot i_{Sc}) - i_{dc} \tag{28}$$

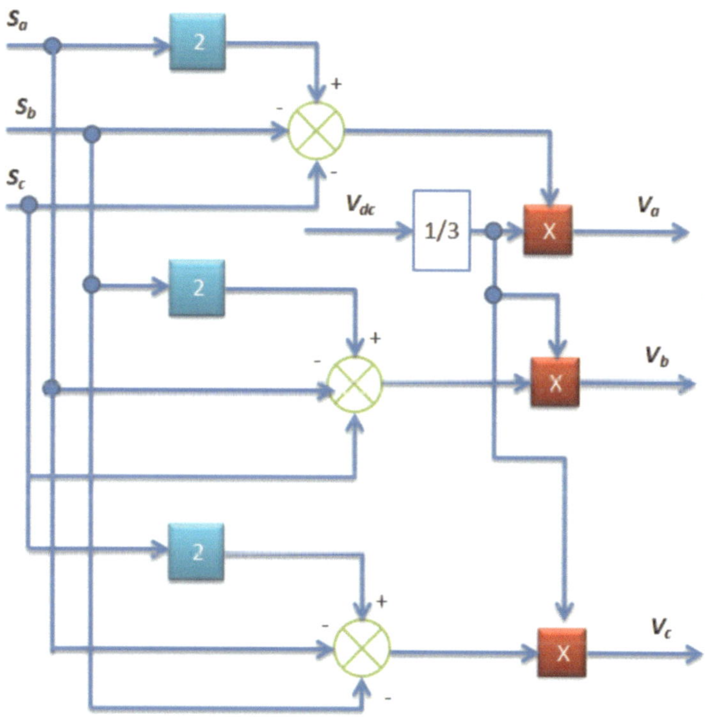

Figure 12.
The block diagram of the RSC.

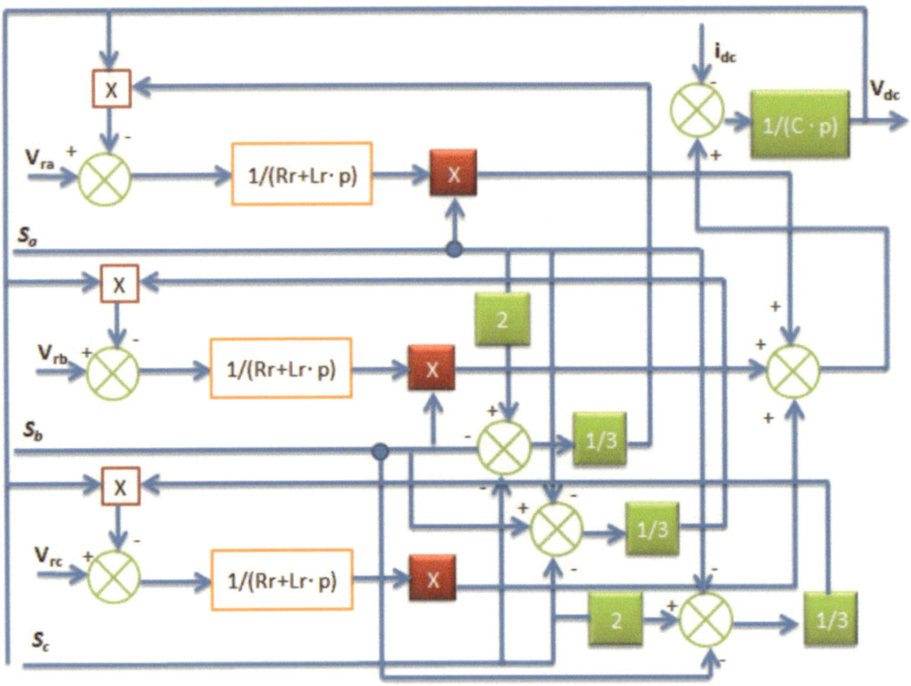

Figure 13.
The block diagram of the GSC.

where V_{rabc} is the three-phase rotor voltage of DFIG [V]; i_{rabc} is the three-phase rotor current of DFIG [A]; C is the capacitor constant [F]; V_{dc} is the DC bus voltage [V]; i_{dc} is the DC output current [A].

The block diagrams of the rotor side converter (RSC) and the grid side converter (GSC) are given, respectively, in **Figures 12** and **13**.

4. Simulation results

The variable speed wind turbine model based on DFIG with a power of 3 Kw has been developed and simulated using MATLAB/Simulink software. The

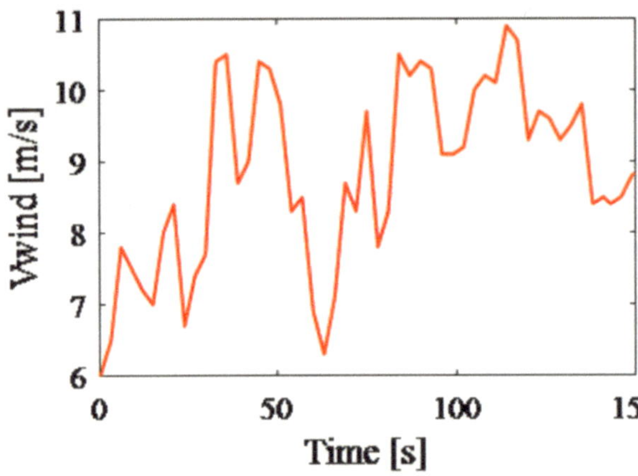

Figure 14.
Wind speed profile [m/s].

turbine and DFIG parameters are extracted from [30]. Some simulation results of the wind system modeled in this study are presented in the figures below. **Figure 14** shows the random wind speed profile applied to the turbine.

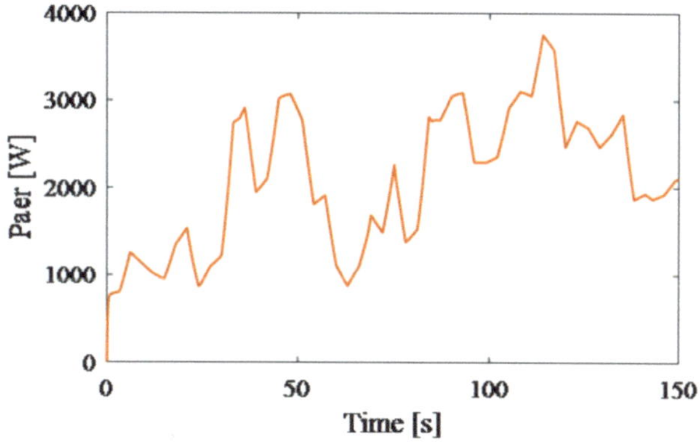

Figure 15.
Aerodynamic power [W].

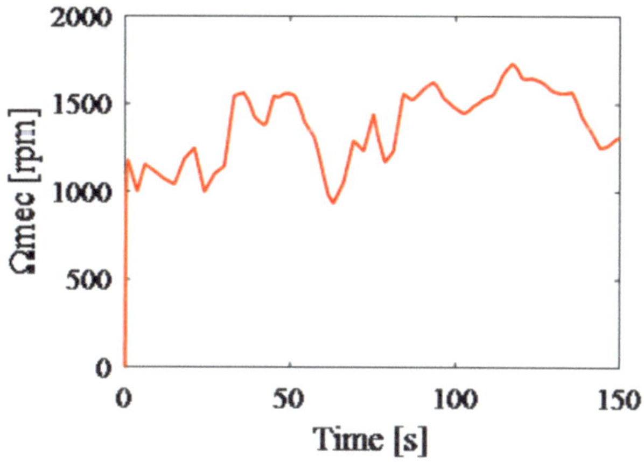

Figure 16.
Mechanical speed of DFIG [rpm].

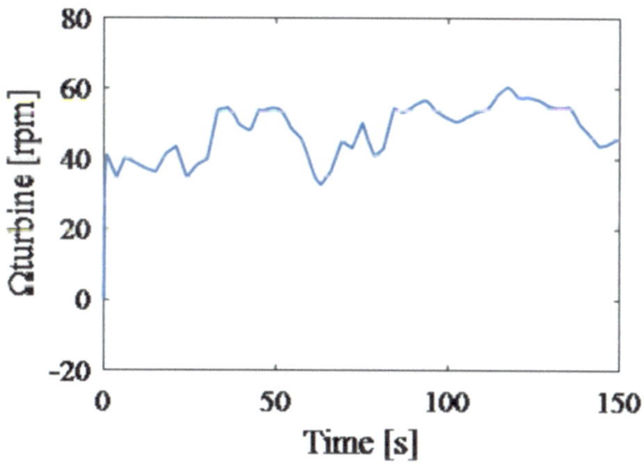

Figure 17.
Turbine speed [rpm].

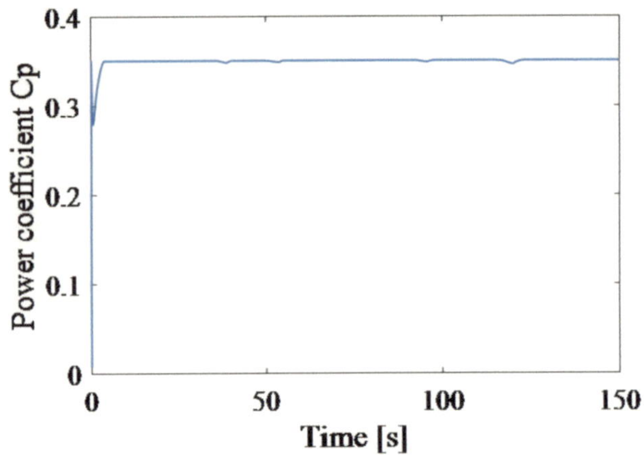

Figure 18.
Power coefficient Cp.

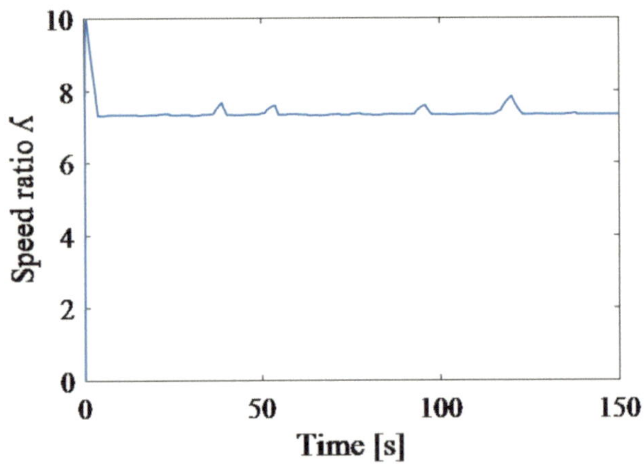

Figure 19.
Speed ratio ʎ.

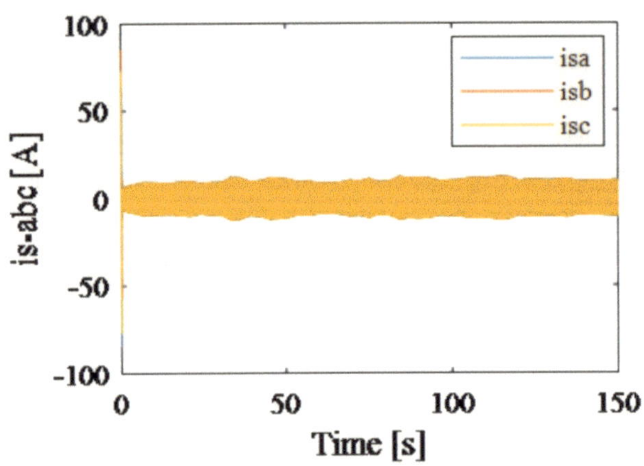

Figure 20.
Three-phase stator current [A].

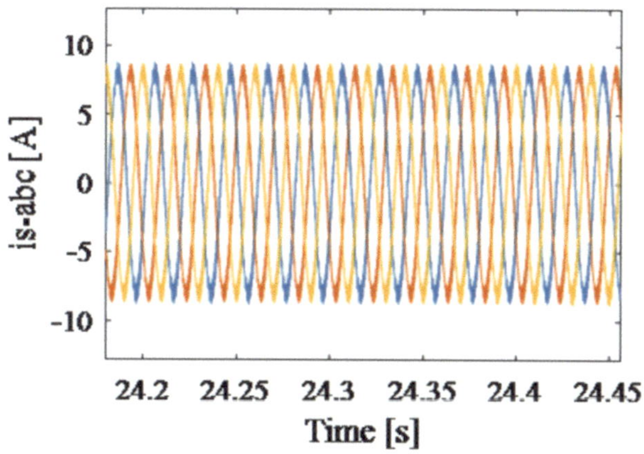

Figure 21.
Zoom on the stator currents [A].

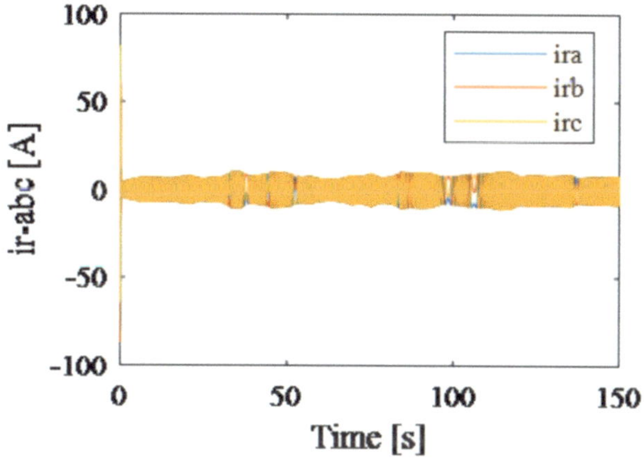

Figure 22.
Three-phase rotor current of DFIG [A].

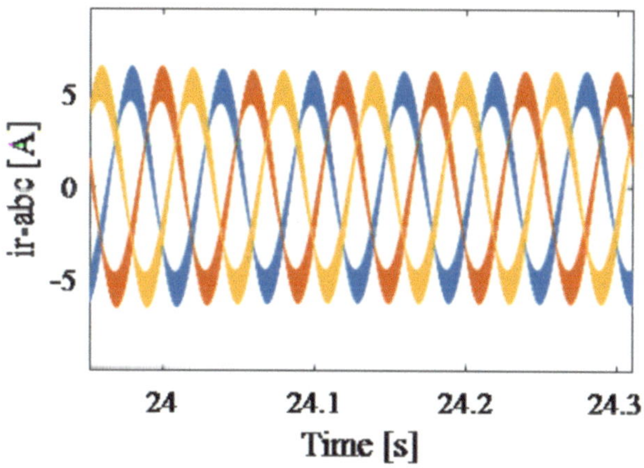

Figure 23.
Zoom on the rotor currents [A].

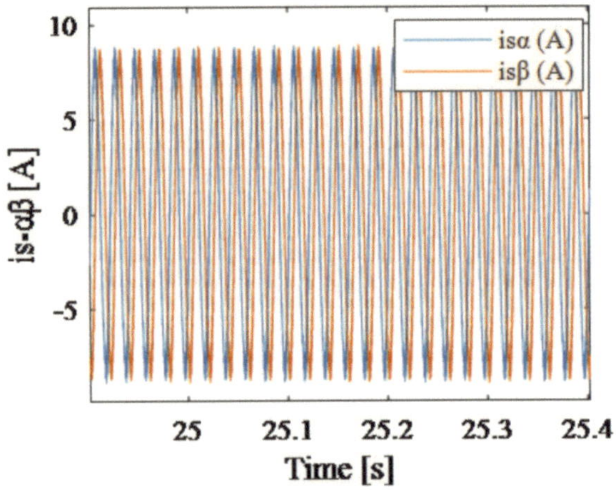

Figure 24.
The stator currents in the (α, β) reference [A].

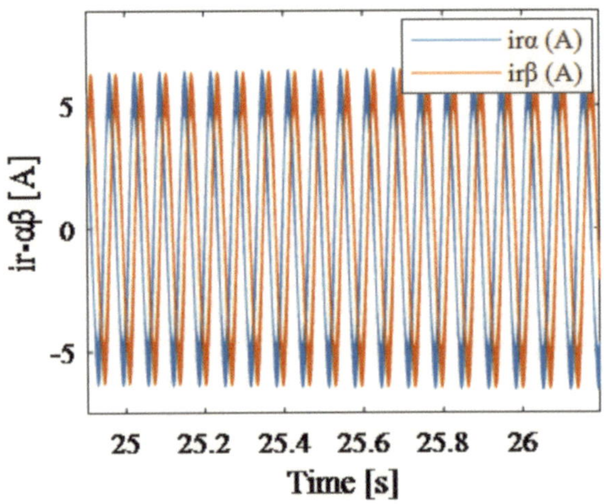

Figure 25.
The rotor currents in the (α, β) reference [A].

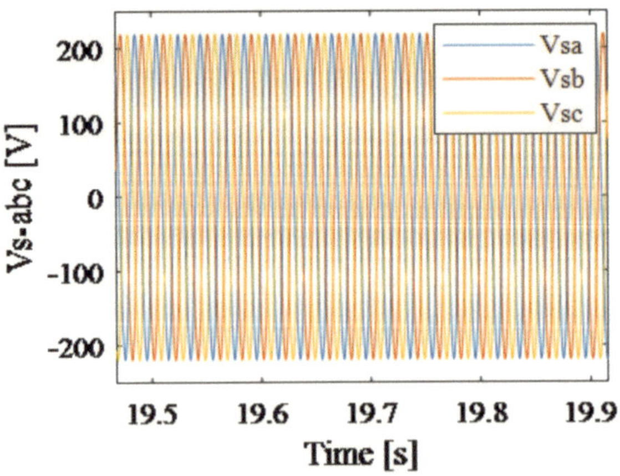

Figure 26.
Three-phase stator voltage [V].

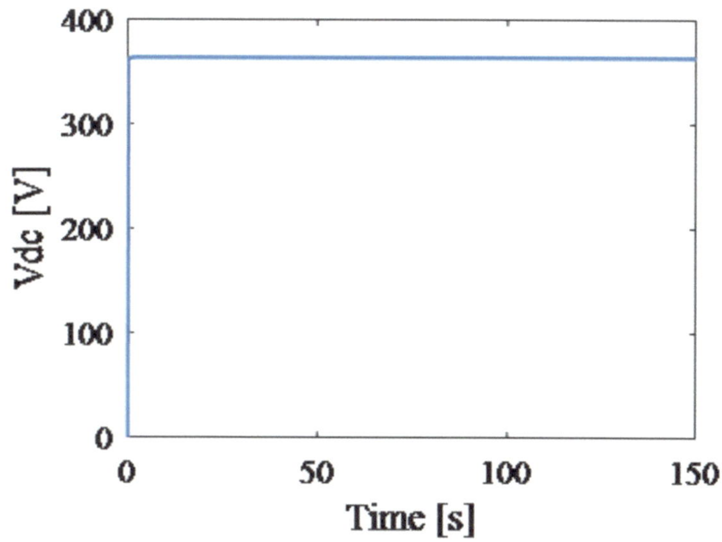

Figure 27.
Voltage of the DC capacitor [V].

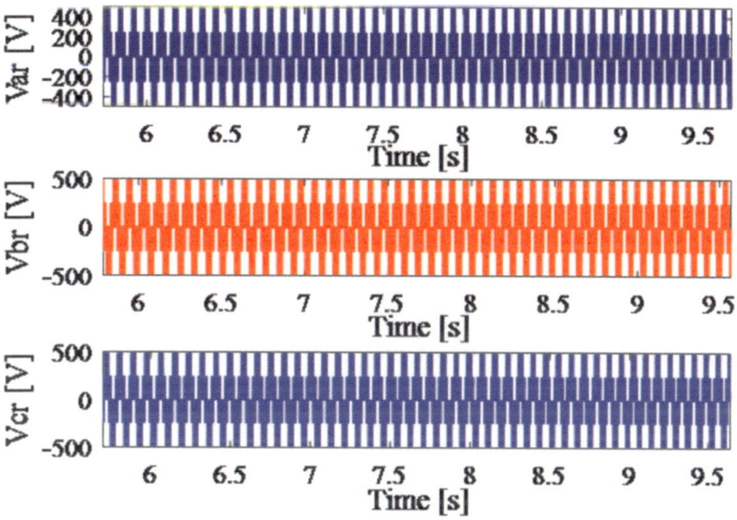

Figure 28.
The output voltages of the rotor side converter [V].

The wind speed varies between [6 m/s] and [11 m/s]. **Figure 15** presents the aerodynamic power delivered by the wind turbine and it reached 3kw when the wind speed is up to 11 [m/s]. **Figures 16** and **17** illustrate respectively the mechanical speed of the generator shaft and the speed of the turbine shaft. It can be noticed from the **Figure 16** that during the simulation time (150 seconds), the generator operates in both hypo and hyper synchronous operating modes.

The **Figures 18** and **19** show respectively the variation of the power coefficient Cp and the variation of the speed ratio λ, which coincide with the maximum power coefficient and with the optimal speed ratio.

The simulation results of the wind system electrical part, including the electrical characteristics of the DFIG, the power converters, and the capacitive bus, are presented in **Figures 20–27**.

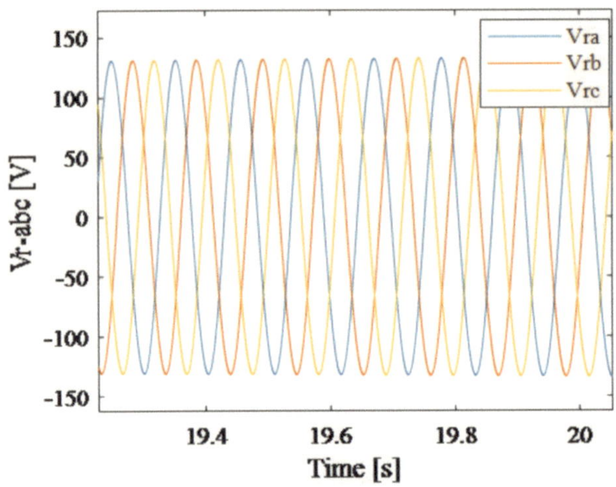

Figure 29.
Three-phase rotor voltage [V].

By applying the first-order passive filter (R, L) to the square-wave signals, given in **Figure 28**, the rotor voltages in sinusoidal form are obtained and shown in **Figure 29**.

5. Conclusion

This chapter presents the modeling and simulation results of the most commonly used speed variable wind turbine driven by a doubly fed induction generator. In order to generate efficient and quick electrical power, the control techniques are applied, such as the MPPT control for wind turbine and the stator flux-oriented vector control are used for the generator. The wind turbine system and its control methods are established in the MATLAB/Simulink environment. The obtained signals are used for the design of fault diagnostic methods in future works.

Author details

Imane Idrissi[1,2]*, Houcine Chafouk[2], Rachid El Bachtiri[3] and Maha Khanfara[1]

1 CED: STI, FST, REEPER GROUP, PERE Laboratory, EST, USMBA University, Fez, Morocco

2 IRSEEM, ESIGELEC, Normandy University/UNIRouen, Rouen, France

3 REEPER GROUP, PERE Laboratory, EST, USMBA University, Fez, Morocco

*Address all correspondence to: imane.idrissi@usmba.ac.ma

References

[1] Global Wind Energy Council, Global Wind Energy Report: Annual Market Update 2017; 2018

[2] Energy Information Administration. Cost and performance characteristics of new generating technologies. Annual Energy Outlook 2018. 2018. pp. 1-3

[3] Local Government Association and The Energy Saving Trust. How much do wind turbines cost and where can I get funding? 2009. Available at: http://www.local.gov.uk/home//journal_content/56/10180/3510194/ARTICLE [Accessed: 11 February 2015]

[4] http://www.ewea.org/wind-energy-basics/faq/

[5] Li H, Chen Z. Overview of different wind generator systems and their comparisons. IET Renewable Power Generation. 2008;**2**(2):123-138

[6] Orabi M, El-Sousy F, Godah H, et al. High-performance induction generator-wind turbine connected to utility grid. In: 26th Annual International Telecommunications Energy Conference, INTELEC 2004. IEEE. 2004. pp. 697-704

[7] Muller S, Deicke M, De Doncker RW. Doubly fed induction generator systems for wind turbines. IEEE Industry Applications Magazine. 2002;**8**(3):26-33

[8] Davigny A. Participation aux services système de fermes d'éoliennes à vitesse variable intégrant du stockage inertiel d'énergie [doctoral thesis]. Doctoral School Sciences for the Engineer, University of Sciences and Technology of Lille; 2007

[9] Belmokhtar K, Doumbia ML, Agbossou K. Modélisation et commande d'un système éolien à base de machine asynchrone à double alimentation pour la fourniture de puissances au réseau électrique. In: Quatrième Conférence Internationale sur le Génie Electrique CIGE. 2010. pp. 03-04

[10] Burton T, Jenkins N, Sharpe D, et al. Wind Energy Handbook. Chicester, United Kingdom: John Wiley & Sons; 2011

[11] Kerboua A, Abid M. Hybrid fuzzy sliding mode control of a doubly-fed induction generator speed in wind turbines. Journal of Power Technologies. 2015;**95**(2):126-133

[12] Petru T, Thiringer T. Modeling of wind turbines for power system studies. IEEE Transactions on Power Systems. 2002;**17**(4):1132-1139

[13] Junyent-Ferré A, Gomis-Bellmunt O, Sumper A, et al. Modeling and control of the doubly fed induction generator wind turbine. Simulation Modeling Practice and Theory. 2010; **18**(9):1365-1381

[14] Luis AS, Wen Y, de Jesus RJ. Modeling and control of wind turbine. Mathematical Problems in Engineering. 2013;**2013**:1-13

[15] Prajapat GP, Senroy N, Kar IN. Wind turbine structural modeling consideration for dynamic studies of DFIG based system. IEEE Transactions on Sustainable Energy. 2017;**8**(4): 1463-1472

[16] Syahputra R, Soesanti I. Modeling of wind power plant with doubly-fed induction generator. Journal of Electrical Technology. 2017;**1**(3):126-134

[17] Widanagama A, Lidula N, Rajapakse AD, Muthumuni D. Implementation, comparison and application of an average simulation model of a wind turbine driven doubly fed induction generator. Energies. 2017; **10**(11):1726

[18] Boukhamkham H. Diagnostique des défaillances dans une machine asynchrone utilisée dans une chaine éolienne [doctoral thesis]. University of Mohamed Khider Biskra; 2011

[19] Toual B. Modélisation et commande floue optimisée d'une génératrice à double alimentation, application à un système éolien à vitesse variable [doctoral thesis]. University of Batna 2; 2010

[20] Azzouz T. Modélisation et commande d'un système de conversion d'énergie éolienne à base d'une MADA [doctoral thesis]. Mohamed Khider-Biskra University; 2015

[21] Tameghe T, Andy T. Modélisation et simulation d'un système de jumelage éolien-diesel alimentant une charge locale [doctoral thesis]. University of Quebec in Abitibi-Témiscamingue; 2012

[22] Report ReGrid: Basics of wind energy. Renewables Academy (RENAC) AG, Schönhauser Allee 10-11, 10119 Berlin (Germany)

[23] Aguglia D, Viarouge P, Wamkeue R, et al. Determination of fault operation dynamical constraints for the design of wind turbine DFIG drives. Mathematics and Computers in Simulation. 2010; **81**(2):252-262

[24] Ackermann T. Wind Power in Power Systems. John Wiley & Sons; 2005

[25] Bechouche A. Utilisation des techniques avancées pour l'observation et la commande d'une machine asynchrone: application à une éolienne [doctoral thesis]. Mouloud Mammeri University; 2013

[26] Atoui I. Contribution Au Diagnostic De Defauts D'une Generatrice Asynchrone Dans Une Chaine De Conversion D'energie Eolienne [doctoral thesis]. Badji Mokhtar University of Annaba; 2015

[27] Sylla AM. Modélisation d'un émulateur éolien à base de machine asynchrone à double alimentation [doctoral thesis]. University of Quebec at Trois-Rivières; 2013

[28] El Aimani S. Modélisation des différentes technologies d'éoliennes intégrées dans un réseau de moyenne tension [doctoral thesis]. Central School of Lille; 2004

[29] Hacil M. Amélioration des performances des énergies éoliennes; 2012

[30] Pascal K. Modélisation et mise en œuvre d'une chaine de production éolienne à base de la MADA. Autre. 2013

[31] Poitiers F. Etude et commande de génératrices asynchrones pour l'utilisation de l'énergie éolienne-machine asynchrone a cage autonome-machine asynchrone à double alimentation reliée au réseau [doctoral thesis]. University of Nantes; 2003

[32] Bossoufi B, Karim M, Lagrioui A, et al. Observer backstepping control of DFIG-generators for wind turbines variable-speed: FPGA-based implementation. Renewable Energy. 2015;**81**:903-917

[33] Idrissi I, Chafouk H, et al. A bank of Kalman filters for current sensors faults detection and isolation of DFIG for wind turbine. In: 2017 International Renewable and Sustainable Energy Conference (IRSEC). IEEE. 2017. pp. 1-6

[34] Shao S, Abdi E, Barati F, et al. Stator-flux-oriented vector control for brushless doubly fed induction generator. IEEE Transactions on Industrial Electronics. 2009;**56**(10): 4220-4228